To Andy

Thank you for everything my friend

Alun.

▶ Technology, Self-Fashioning and Politeness in
Eighteenth-Century Britain

DOI: 10.1057/9781137467485.0001

Also by Alun Withey

PHYSICK AND THE FAMILY: Health, Medicine and Care in Wales, 1600–1750 (Manchester: Manchester University Press, 2011).

* Winner of European Association for the History of Medicine and Health (EAHMH) book prize, 2013

DOI: 10.1057/9781137467485.0001

palgrave▸pivot

Technology, Self-Fashioning and Politeness in Eighteenth-Century Britain: Refined Bodies

Alun Withey

Wellcome Research Fellow, University of Exeter, UK

palgrave
macmillan

DOI: 10.1057/9781137467485.0001

First published 2016 by
PALGRAVE MACMILLAN

The author has asserted their right to be identified as the author of this work
in accordance with the Copyright, Designs and Patents Act 1988.

Palgrave Macmillan in the UK is an imprint of Macmillan Publishers Limited,
registered in England, company number 785998, of Houndmills, Basingstoke,
Hampshire, RG21 6XS.

Palgrave Macmillan in the US is a division of Nature America, Inc.,
One New York Plaza, Suite 4500 New York, NY 10004-1562.

Palgrave Macmillan is the global academic imprint of the above companies
and has companies and representatives throughout the world.

Hardback ISBN: 978-1-137-46747-8
E-PUB ISBN: 978-1-137-46749-2
E-PDF ISBN: 978-1-137-46748-5
DOI: 10.1057/9781137467485

Distribution in the UK, Europe and the rest of the world is by Palgrave
Macmillan®, a division of Macmillan Publishers Limited, registered in
England, company number 785998, of Houndmills, Basingstoke,
Hampshire RG21 6XS.

Library of Congress Cataloging-in-Publication Data is available from the
Library of Congress

A catalog record for this book is available from the Library of Congress

A catalogue record for the book is available from the British Library

Contents

DOI: 10.1057/9781137467485.0001

List of Figures

DOI: 10.1057/9781137467485.0002

Preface

In the eighteenth century, expectations of bodily appearance changed. Cultural changes, including the rise of 'polite' behaviours and deportment, meant that managing the body was increasingly requisite in order to meet changing social expectations of appearance. In came new aesthetic ideals including smooth skin, neatness and harmony, symmetry, form and concepts of a 'natural' body. An increasing number of arenas, from coffee houses to civic and society events also brought new opportunities for bodily appearance to be scrutinised.

To cater for changing expectations of bodily appearance, a range of instruments, objects and practices became invested with new meaning. All manner of devices were used to alter everything from the minutiae of self-fashioning to the very shape and form of the corporeal fabric. Razors helped men achieve the clean-shaven masculine ideal. Tweezers, nail nippers and toothpicks all played an important role in cleaning, honing and perfecting bodily surfaces. A variety of machines and instruments helped adults and children mould their bodies towards a 'natural' shape, and to correct and conceal deformities or disabilities.

At the same time, technological developments in the second half of the eighteenth century, most notably in steel, brought changes to what was materially possible in terms of fashioning the body. In many ways it was arguably *the* enlightened metal, and almost certainly the industrial output with which people had the most intimate physical contact. Steel, for example, transformed the manufacture of

DOI: 10.1057/9781137467485.0003

razors, which became sharper, more durable and arguably more aestheti-cally appealing. The tensile, springy strength of cast steel was utilised by specialist makers of postural devices to force the body into the desired shape. It was a key component in the majority of devices discussed in this book, and makers often trumpeted its use in their products.

Despite the apparent emphasis upon innovation and the impact of a new technology, this study is not a straightforward story of change nor, especially, of triumphant progress. At its heart, however, it is a story of refinement and improvement in the broadest sense. Each chapter explores a different instrument, or set of instruments, and demonstrates the often-intricate relationship that existed between form and function, technology and culture. The pace of change differed in each case, and various instruments came to prominence or popularity at different times. Changes to spectacles, for example, were discernible before 1750 and continued beyond 1800. The razor-making trade began to be trans-formed around the mid-century. Developments in surgical instrument manufacture and retail, however, were more prominent at the end of the century. Most instruments discussed here also follow a different trajec-tory in terms of the bodily expectations to which they were applied. Razors spoke to masculine ideals of the smooth, clean-shaven face. Personal grooming instruments, such as toothpicks and nail nippers, were a response to new emphases laid upon the appearance of specific areas of the body in public. Further, the relationship between form and function is often subtle, and the interplay between technology and culture difficult to separate. Steel, for example, allowed changes to the design of spectacles, but this occurred at the same time as they began to become associated with learning and sagacity, rather than deficiency. Even surgical instruments, to some degree transformed by steel, were not part of what might be viewed as domestic consumption, but they were nevertheless important vectors for changes to bodily form.

In the last analysis, all objects and practices are historically contingent. The eighteenth century was a period in which all manner of devices were invested with new meaning as vectors through which people sought to manage, refine and even transform, their appearance. In many cases, these were quotidian objects rather than trophy possessions like clothing, watches or jewellery. If we are to capture the totality of experiences, it is necessary to engage with the everyday, the quotidian and the mundane as well as the elaborate and luxurious. It is to these ends that this book aspires.

DOI: 10.1057/9781137467485.0003

Acknowledgements

The creation of this book owes much to the countless contributions made by colleagues and friends over the past few years. Perhaps the biggest debt of gratitude is owed to Chris Evans, who first inspired my interest in the history of steel during my tenure as research fellow on his project 'Steel in Britain in the Age of Enlightenment' at the University of South Wales. His generosity in allowing me to appropriate part of the material from that project is matched by his continuing support, advice and encouragement for my work.

The book was written whilst I have been a Research Fellow at Exeter University, working on the Wellcome Trust project 'The Medical World of England, Wales and Ireland, 1550–1715', under the guidance and supervision of Jonathan Barry and Peter Elmer. It is a tribute to their patience, forbearance and willingness to invest in the career development of their team that this book could be written at all. Jonathan has also read and provided incisive and valuable comments upon many draft chapters, which have helped immeasurably in shaping the book, for which I am extremely grateful. My colleagues Justin Colson, John Cunningham and Hannah Murphy all provided friendly and helpful advice in our many meetings. I also particularly thank Margaret Pelling for encouraging me to complete the book, and giving me the confidence to do so when it looked in danger of losing direction. The History Department at Exeter and the Centre for the History of Medicine have provided a friendly, collegial

DOI: 10.1057/9781137467485.0004

and supportive atmosphere in which to work, and I am proud to be their colleague.

Throughout my academic career I have also been blessed with many good friends in the academic community who have provided encouragement, tea and sympathy as well as offering advice, suggesting source material, and commenting on drafts. Andy Croll has been a steadfast friend and mentor since my postgraduate days, and the book owes much to his sound advice, sympathetic ear and many much-needed confidence boosts. David Turner, my friend and former PhD supervisor, has likewise been a constant advocate, whose work on the history of disability also inspired my interest in the technology of postural devices.

Neil Handley, curator of the British Optical Association Museum in London, provided much help and advice over the past few years, and I am grateful for his generosity in providing much source material relating to the history of eyewear.

Will Pooley and the audience at the 'History of the Body: Approaches and Directions' colloquium at the Institute for Historical Research in May 2015 offered extremely useful and thought-provoking comments on a paper I delivered there, based on the themes of this book. Karen Harvey and Michele Cohen provided help and guidance relating to the chapter on razors, and I am also grateful to Matthew Grenby for permission to adapt part of my 2012 article in the *Journal for Eighteenth-Century Studies* for the same chapter.

I would also like to thank Jonathan Bohen, Isobel Carr, Helen Ross, Annie Gray, Thony Christie and Sara Read for their generosity in providing source materials for various chapters in the book.

Finally, the book could not have been written without the constant love and support of my family. To my wife, Suzanne, son, Morgan and mother, Pat, who have accompanied every step of this journey, and probably feel they know the eighteenth-century body as much I do by now, my love and grateful thanks.

DOI: 10.1057/9781137467485.0004

Introduction: Framing the Enlightened Body

Abstract: *The eighteenth century brought changes in attitudes towards shaping the body. Previously, bodily shape and form, including disability, were viewed as God's work, and beyond the hand of man. Equally, treating deformities and visible impairments had once exemplified vanity and pride. But a new culture emerged which made conquering the body a noble and justifiable act. At the same time, technological improvements, notably in steel, yielded a huge range of products for the body, encompassing everything from large apparatuses for altering body shape, to the smallest instruments of personal grooming. While 'politeness' is often seen as a system of language and behaviours, Withey suggests that the body itself could be 'polite'. Arguing that steel instruments were key enablers in this process, Withey provides an important new study of the complex relationship between technology and the body.*

Withey, Alun. *Technology, Self-Fashioning and Politeness in Eighteenth-Century Britain: Refined Bodies.* Basingstoke: Palgrave Macmillan, 2016. DOI: 10.1057/9781137467485.0005.

This book explores the ways in which various technologies of the body in the eighteenth century were refined by the use of a new type of steel – cast, or crucible, steel. In so doing it also explores the important relationship between bodies and instruments during the Enlightenment. Steel was itself the product of a continual process of refinement and innovation in metallurgy during the second half of the eighteenth century. Many of these products offered new possibilities for the material alteration of the body and its surfaces, at the same time as concepts of the body were in flux, with changing attitudes towards everything from gender to deformity. Refinement, however, does not imply that everything in this book is entirely new. Instead, new technological developments enabled significant changes and (in some cases) improvements to existing instruments for the body.

The eighteenth century was a period of flux in attitudes towards the body, appearance and gender. Previously, the impaired body had been viewed as a fait accompli, its owner condemned forever to endure whatever deformities God or Nature had bestowed upon it. In the early part of the century, debates raged about the dangers of vanity as well as the morality of trying to interfere with God's work. Nevertheless, by the mid-eighteenth century, there were changes in attitudes. If the management of appearance, including treatment of deformities and visible impairments, had once exemplified vanity and pride, new, enlightened themes such as 'improvement', self-control and mastery began to make conquering the body a noble and justifiable endeavour.[1] In addition, attention was increasingly being paid to the minutiae of bodily surfaces, like faces and hands, which were seen as holding the key to inner characteristics.

At the same time as these broader social and cultural changes were taking place, a febrile environment of artisanal experimentation in metallurgy, especially in steel, and product innovation across medical and scientific trades yielded a huge range of 'technologies of the body'. These encompassed everything from large apparatuses for altering bodily shape, posture and gait, to the smallest, quotidian items of personal grooming such as tweezers and nail nippers. In some cases, new technologies transformed the design of instruments, while in others, the instruments themselves took on important new meanings as vectors through which individuals could aspire to changing ideals of the body.

In London, for example, specialist makers of devices for the body took to the advertising pages of the popular press to highlight their wares. The truss maker J. Eddy of Dean Street, Soho, advertised many products for

DOI: 10.1057/9781137467485.0005

the body, including rupture trusses, instruments to 'cure deformities in the body and legs', elastic bandages and leg irons to 'cure' deformity. Eddy claimed that his products were designed to 'relieve' the 'various complaints and infirmities not in the power of medicine or surgery'. Another truss maker, George Dowling of Duke's Court, advertised similar items, but also went further in offering devices specifically tailored towards correcting, concealing or improving posture. These included 'various sorts of instruments to redress distorted legs, bow-legs, knock knees, etc' as well as '[l]adies backs and collars [a] machine for the distorted spine' and others. Dowling stressed that '[b]odily Deformities [could be] remedied in appearance, so as not to be distinguished from perfect nature'.[2]

But it was not just specialists like Eddy and Dowling who sold products to alter or refine the body. A mere stone's throw from their premises in Soho and Duke Street was the shop of the cutler John Brailsford. His trade card listed his multifarious stock from cutlery and scissors to penknives and spurs, along with rough drawings depicting some of his goods. At first glance they illustrate well the stock in trade of a cutler. But closer inspection reveals some perhaps more surprising items, many of which relate to the body, including combs, brushes, razors, tweezers, scissors and even pince-nez spectacles. Brailsford's card demonstrates both the ubiquity of products for the body for sale in the eighteenth century and the diverse retailers selling them.

Eddy, Dowling and Brailsford, along with a raft of others, catered for a developing market for technologies of the body, especially after 1750. Their advertisements capture a number of important concepts. First, it was possible to change the shape or cosmetic appearance of the human body and, as Eddy's advertisement stressed, this could be done without recourse to medicine or surgery. Second, concepts of the body were in flux, and many devices, instruments and machines were available to help people meet changing expectations and ideals of form. Within this group, instruments made of steel played an important role as enablers of bodily change. It is these products, their design, marketing, consumption and meanings with which this book is concerned.

Technologies of the body

The place of new technologies in an 'industrial enlightenment' has begun to be explored. As Celina Fox argues, industrial processes and premises

DOI: 10.1057/9781137467485.0005

became increasingly interesting both to a scientific elite, concerned with classifying and ordering the natural world, and to a public beguiled by notions of scientific advancement and progress.[3] Leading industrialists articulated enlightened values through product innovation and advertising of 'new' goods, and even by opening up their premises to curious savants as part of the 'Grand Tour'.[4] And yet, historians have largely overlooked questions of how such 'enlightened' industry and manufacturing shaped individuals' concepts of politeness. Technological innovations around the 1740s made steel an increasingly abundant and important good, but also a component in the fashioning of a new, refined self. While crucible (or cast) steel is understood as an innovative industrial process, its non-industrial uses are seldom considered. Yet steel was vital for some of the most personal rituals of everyday life.[5] It was the metal with which people had the closest, even the most intimate, physical contact.

In 1700, the most common type of steel in Britain was obtained through the cementation process.[6] Bars of wrought iron were packed in stone sarcophagi in alternate layers with charcoal, and heated in a furnace. Although it did not melt, the iron absorbed carbon from the charcoal through the intense heat, causing the characteristic markings giving 'blister steel' its name.[7] Blister steel was uneven in quality due to the uneven distribution of carbon. A more uniform and homogenous steel was obtainable by reheating and hammer-forging bars of blister steel together to form a single ingot of 'shear steel'. Although more uniform than blister, it suffered from the same heterogeneous deficiencies and was therefore unsatisfactory for use in edged tools. Complete uniformity could only be obtained by manually removing slag impurities in a crucible, but conventional charcoal furnaces could not generate the extremely high temperatures required. Through his development of a new type of coke furnace and repeated experimentation over a number of years, Benjamin Huntsman's successful melting of blister steel was the genesis of cast steel in the late 1740s.

Steel's physical properties rendered it signally useful as a material component in construction. Yet a commodity so closely intertwined with earthy industrial labour also became the acme of Enlightenment taste and fashion. The jewellery market, for example, was certainly transformed by steel. Diamonds were the height of luxurious and conspicuous consumption, and costume jewellery reflected social mores related to society ritual and appearance. They were also prohibitively expensive, severely limiting the market.[8] Steel, however, offered new possibilities as an ersatz

DOI: 10.1057/9781137467485.0005

precious metal. Here was a material that offered all the decorative allure of diamonds. Cut and faceted into imitation stones called 'brilliants', cast steel coruscated. With flat surfaces polished, it shone mirror-like. Fashionable gentlemen embraced cast steel watch chains, both to support their newly modish gold and silver watches and also as costume adornments in their own right.[9] Other accoutrements like seals and lockets drew attention to the means of the wearer. In the 1760s, chatelaines made from 'blued steel' presented a 'gamut of metallic hues', while glistening steel buttons were the coming thing for the well-dressed man.[10]

Historians have begun to explore the complex relationship between bodies, machines, instruments and mechanisms in the eighteenth century, and their effect upon appearance, image and identity. As Michael Polanyi noted, eighteenth-century scientific instruments could become 'an extension of the senses'.[11] A growing literature explores the nature of the disabled body and the emergent market for devices to 'correct' bodily deformity. Liliane Perez and Christelle Rabier have analysed the nascent trade in rupture trusses in the eighteenth century within the broader market for 'toys', and their close relationship to emerging metallurgical technologies.[12]

Postural devices, for example, were part of a large range of bodily technologies that either penetrated the skin or were applied to its surface.[13] Lynne Sorge-English has highlighted the importance of the stay in both 'correcting' deformities and shaping an ideal female form, and the role of clothing in general as a means to shape the body.[14] In tandem, a broad literature on the medical marketplace has explored the role of bodily technologies and instruments, including paramedical devices from 'metallic tractors' to rupture trusses.[15] As retailers and manufacturers adopted a genteel language of advertising, corrective devices began to lose their associations with defectiveness and deformity, and became subsumed within the broader culture of consumer demand for polite goods. Steel instruments were an important component in this process.

The physical properties of cast steel let people fashion their bodies in new ways, and in turn reflect shifting ideas about desirable bodily shape and form. Posture, for example, became increasingly important as people sought to 'correct' poor stance or deformity, and seek a body that was straight and erect. The tensile, springy strength of steel rendered it a key component in this process. Steel devices, including collars and neck swings, were marketed to help people achieve this without resorting to a practitioner.

DOI: 10.1057/9781137467485.0005

Just as bodily shape was being refined, so bodily surfaces were also sites of transformation. Steel razors allowed men to reflect new ideals of the masculine face as being smooth, elegant and refined. Manufacturers played upon the technological innovation involved in crafting their new blades. Accompanying the refinement of bodily shape was a new focus upon personal grooming. Authors such as Nicholas Andry set out ideals of size and proportion for bodily limbs, organs and features. As new emphasis was placed upon physical features and expressions as indicators of sensibility and politeness, paying attention to one's appearance, especially the face and hands, became important.[16] Tweezers, nail nippers and, to some extent, razors were also part of the broader trade in 'toys'. Steel toys and trinkets, along with various kinds of instruments, were part of a 'non-verbal language' of show and display. Being both decorous and functional, what Liliane Perez terms 'artful mechanisms' had a central role to play in self-fashioning.[17] Steel, and other metals, was vital in this process. And yet, virtually no attention has been paid to the role or importance of the quotidian instruments used in this process.

Spectacles, once viewed as symbols of deficiency, took on new meanings as exemplars of learning and sagacity. Again, steel played an important role in this transformation, both in functional and aesthetic terms. Its elastic properties enabled changes to be made to spectacles' sides (arms), letting them adhere to the wearer's head. The aesthetic appeal of polished steel spectacles also transformed them from items to be hidden, to something to be shown off. The eye was privileged as a key organ of sense and as the symbol of both literal and philosophical vision. The rise of print and the popularity of reading, together with the desirability of steel as a fashionable accessory all combined to bring about changes to conceptions of visual aids, their form, function and marketing. Of course, spectacles were also optical instruments, affected by new developments in glass- and lens making, and as such often produced or sold by instrument makers, but this should not obscure the importance of their frames. As today, these were often the focus of their marketing.

'Technologies of the body' were, therefore, integral components of the burgeoning marketplace of Enlightenment England. All manner of bodily devices and instruments became available and were advertised for people to use upon themselves. Many areas of the body saw specialist makers and practitioners, from truss makers to optometrists to chiropodists, who often sold products on the basis that an individual could apply them without recourse to a practitioner. Indeed, consumers were addressed as 'customers' rather than 'patients', and were served by makers claiming to succeed where medicine had failed: as such, they occupied a liminal space between the medical marketplace and the markets for scientific instruments and toys.[18] Even surgical instruments, whilst clearly not used by individuals upon themselves, were important in transforming the body and were shaped in complex ways by the introduction of cast steel.

Polite bodies?

To understand the role of these instruments in transforming the body, however, it is necessary to explore the changing climate of ideas surrounding the body.

Historians of the eighteenth century have begun to explore how new manufactured goods helped in the construction of a new type of 'self'. 'Polite' manners and behaviours were entwined with ownership of the right goods, the wearing of the right clothes and attendance at the right social events. As Lawrence Klein has noted, 'politeness' has been revived as an emblem of the enlightened age.[19] Nonetheless, the question of how this rather nebulous concept was understood and expressed at an individual level is complicated, not least by the multiplicity of meanings attached to the term itself.

For John Brewer, politeness was a system of social actions, encompassing language, behaviour, gesture and the regulation of mind and body.[20] Implicit within Klein's definition of politeness is also the importance of 'form' and simplicity.[21] Politeness has been described as an 'eighteenth-century idiom' and a synonym for other terms such as 'civil', 'genteel' and 'mannered', but also a term the meanings of which shifted greatly according to context and through time.[22] Whilst conversation, education and manners dominated early conceptions of polite behaviours, appearance and form also became increasingly important.[23] In studying the embodiment of politeness and the ways in which adornments in particular could contribute to the construction of a polite bodily image, increasing attention has been paid to the impact of enlightened thinking upon the individual body. Historians of gender and masculinity, for example, have explored the notion of a 'polite gentlemanliness', highlighting the importance of rules of conduct and behaviour alongside status indicators, from wigs to couture, and from watches to jewellery.[24] Karen Harvey

DOI: 10.1057/9781137467485.0005

and others have noted the relationship between embodiment and the domestic interior in the eighteenth century.[25] Others highlight the role of advertising and the importance of 'polite consumption' as vectors of the enlightened self.[26] As Richard Sennett has suggested, the eighteenth-century body was a mannequin upon which were hung conventions of fashion, taste and politeness.[27] Newly modish items such as steel jewellery gave material expression to enlightened and polite values, allowing the wearer to conform to public and private expectations of conduct and appearance. Central to this book, however, is the argument that politeness was a concept that extended beyond words, gestures, behaviours and adornments, and encompassed the fabric of the body itself.

European writers considered the body of the 'man of fashion' as fundamentally different in constitution to that of the labourer. In *The Disorders of People of Fashion* (1772), the Swiss physician Samuel Tissot distinguished between the hardy body of the labourer, through its constant exposure to harsh elements, and the somewhat delicate and fey body of the gentleman.[28] If the latter was physically slighter, however, it was also delicate and refined. Contemporaries also acknowledged the difference between mechanical and 'sensible' bodies. In 1749, the English philosopher David Hartley argued for two types of motion to which the body was subject. First were involuntary motions, such as the movements of the heart and the bowels, over which an individual had no control. Second, however, were voluntary motions, which Hartley defined as arising from ideas and affections of the mind. These motions governed actions such as walking and standing, 'when attended to ... *and performed with an express design*' (my emphasis).[29] In other words, thoughts, feelings and fancies shaped bodily actions and postures. If a person were 'polite' of mind, so they could also be polite of body. While contemporaries never referred directly to bodies as being polite, they acknowledged the role and importance of the body in articulating politeness. A 1775 essay on the characteristics of politeness in the *Universal Magazine* argued that it was a holistic concept governing not only 'temper of mind and tenour of conduct' but also '*bodily appearance, posture and mien* (my emphasis).[30] A polite gentleman (the essay addressed men particularly) should embody the posture of a fencer, the gait of a dancer, the ear of a musician and the mind of a philosopher. Such a person 'walks by rules of art, dictated by nature'.[31] Herein lay the key to a polite body. It was a work of art, and therefore an ideal, but one simultaneously governed by the immutable rules of nature. It could be achieved by artificial means, so long as this

DOI: 10.1057/9781137467485.0005

process went no further than restoring or correcting, but not for reasons of artifice, deception or vanity. In fact, as Chapter 1 argues, nature was at the heart of debates about bodily form. Nevertheless, debates surrounded the form of this supposed 'natural' body. Some saw it as one closest to the state of nature, in the bodies of the poor or the inhabitants of far-flung nations whose bodies had been unadulterated with devices. Indeed, some even viewed interference with, or alteration of, the body as inherently unnatural, reinforced by the twisted and bent bodies caused by overzealous use of trusses, bandages and stays. However, much of the book documents attempts to 'correct', conceal or otherwise give the illusion of a 'natural' form – a claim made by the makers of many postural devices, for example.[32]

Some important questions must be raised. First, if there was some understanding of a polite body ideal, how widespread was it? Much evidence for advertising and consumption is based around London. Part of the reason for this is evidential since the majority of newspaper sources are based in the capital. Likewise, as the growing historiography of instrument manufacture has demonstrated, London, along with Sheffield and Birmingham, was a centre for metallurgical production and innovation. Where evidence exists from provincial England, this has been included, and suggests that patterns were broadly repeated outside London.

A more difficult issue is that of the social depth of the consumption of bodily technologies and, by extension, the underlying attitudes informing it. Again, partly as a result of evidence, the focus of this book is necessarily urban and elite. It might well be assumed, as the continuing debates about emulation have done, that people lower down the social scale were equally engaged in the transformation of their bodies. Poor men clearly shaved, since village barbers were commonplace. Evidence presented in the chapter on postural devices suggests that people across society were prepared to intervene to correct or improve their posture. Spectacles were available at a range of prices. The question is whether it was only the relatively narrow section of urban polite society to whom any notion of a polite body applied. Unsatisfactorily, that question remains beyond resolution at present.

The matter is further complicated by debates about the nature of the relationship between politeness and 'sensibility'. The book is largely cast in a period that some historians argue is one of transition from politeness, with its emphasis upon public presentation, to sensibility, which

DOI: 10.1057/9781137467485.0005

encouraged an emphasis on inner sentiment, emotion, melancholy and delicacy of feeling.[33] However, between the 1760s and 1790s, as Philip Carter has argued, the two coexisted, sharing many common factors.[34] The focus here is, in essence, on the importance of public display and the maintenance of the body to suit social ideals. Keeping up bodily appearances remained a vital consideration well into the nineteenth century.

A note of caution must also be struck about the dangers of assuming 'progress' when discussing new technologies. At its heart the book is indeed a story of 'improvement', both in the functional sense and in the contemporary understanding encompassing social, moral and physical refinement. It is, though, equally a study of transformation. Historians are often understandably cautious to avoid triumphalist narratives of science or Whiggish tales of discovery and progress. Cast steel certainly did render improvements to existing instruments and devices in comparison to what had gone before. Sharper razors, for example, made shaving more comfortable. The elastic properties of steel enabled bodies to be forced into the desired shape in a more effective way. The form of surgical instruments from catheters to amputation knives was altered, leading to changes in surgical techniques. In all of the cases discussed in the book, instruments were refined, made sharper, smaller and, more generally, fitter for purpose. In their construction and application, they were 'better' than what had gone before. Nonetheless, it would be cold comfort for the individual about to face having their leg sawn off without anaesthetic to learn that the surgeon was using a new, straight amputation knife! In fact, whether or not these technologies were 'better' is not necessarily the question. Rather, it is the processes of transformation that are of interest here. Many technologies for the body were transformed by cast steel. In turn, people used many these new instruments and devices in order to transform their own bodies. It is the role played by instruments in the purposeful shaping of the body, rather than a story of progress, which the book charts.

Finally, the novel set of sources considered in this book allows us to test assumptions about bodily appearance and conduct, from gendered notions of the body to the chronologies of change. The fact that each chapter follows a different chronological trajectory is useful in gauging both attitudes to bodily ideals and self-presentation. Clearly, in some ways this mix of chronologies confounds a smooth narrative of change, but they offer new perspectives into various important debates in the eighteenth century.

DOI: 10.1057/9781137467485.0005

Structure

Rather, then, than offering a chronological narrative of technological change, the book instead takes a thematic approach, with each chapter focussing upon different instruments or devices. Indeed, as is clear throughout, the introduction of cast steel, around 1750, did not effect an instantaneous transformation across all products. Many of the instruments discussed here had a history before cast steel, but were 'improved' or transformed by it over time. Cast steel razors, for example, were sharper, more durable and more comfortable to use than their predecessors, but did not reach their apogee until the 1770s. The design of spectacles began to alter earlier than razors, but the associated cultural changes took longer to become embedded, lasting into the nineteenth century. It should not therefore be assumed, and indeed is not argued here, that cast steel simply swept away all before it. Neither is it argued that it was the only material for instruments of the body. It was, however, an enabling material the properties of which rendered it uniquely suitable across a broad range of products for the body. That is the organising principle governing the selection of instruments here.

The first chapter explores the marketplace for postural devices. The eighteenth-century body was a site of dispute over what represented the perfect human form. Visible deformity and disability were not only uncomfortable to the sufferer but also carried pejorative connotations that left the 'crooked' open to ridicule. Marked bodies could be socially limiting and bore negative associations. If there was an ideal human form, it was generally straight, erect and symmetrical. Various medical and lay authors attempted to define aesthetic ideals of the human form or to lay down rules for symmetry and form. Some, such as the artist William Hogarth, made a case for the importance of the curve as seen in nature. Whilst the treatment of hernias had spurred the development of various elastic and steel trusses, the period also witnessed a burgeoning market for devices to improve posture. These included items worn within or underneath clothing, like back 'monitors', steel collars and stays, to larger apparatuses such as 'screw chairs' and 'neck swings'. A primary audience for these devices was parents, keen to mould the bodies of their offspring into an acceptable form. Adults also proved willing to intervene in the shaping of their own bodies. As advertisements from postural device manufacturers attest, a new domestic market was emerging, which encouraged individuals to 'treat' themselves without recourse to

DOI: 10.1057/9781137467485.0005

a medical practitioner. One of the key selling points of such devices was their ability to be discreet and 'indistinguishable from nature'. Debates also raged about the moral rectitude of shaping the body in general, but also of the pain and discomfort inflicted upon children in particular, in the name of fashion.

Chapter 2 discusses the impact of cast steel upon razors, amongst the almost total disappearance of facial hair from the male face in the eighteenth century. The capacity of steel to carry a fine edge enabled the production of new, sharper, more durable and even more aesthetically pleasing razors. Sharing common ancestry with surgical-instrument makers, razor manufacturers stood at the very forefront of technological innovation. They continually experimented with steel, using this metallurgical expertise to diversify into other trades. Their specialist knowledge rendered them useful arbiters of quality for steel producers, while their experience fed back into a continuum of metallurgical knowledge. Where once the barber had been the sole provider of shaving services, men increasingly began to shave themselves. Razor makers took advantage of newspaper advertising space to puff their new products, using the language and imagery of polite consumption, but also foregrounding their metallurgical expertise. Using cast steel in razors became a selling point, bolstered by references to the scientific and philosophical credentials of the manufacturer.

The third chapter follows a different line in exploring the role of instruments in personal grooming. Whilst attention has focussed on hygiene and cosmetics, including contemporary debates about make-up and the concealment of the true face, the instruments used by people to refine their appearance are seldom considered. And yet these were vital in the daily management of the self. Unlike other devices discussed here, however, small instruments like nail clippers, tweezers and toothpicks were seldom – if ever – sold on their own account. Whilst steel certainly altered design and function to some extent, the effect was subtler than elsewhere. These items were manufactured not by specialist makers, but by 'toymen' and tool makers. They were not advertised or marketed individually, but were instead part of toilette or equipage kits, which were often elaborate and decorous, belying their quotidian function. But personal grooming grew in importance in the broader context of the enlightenment obsession with the body beautiful. As increasing attention focussed on the minutiae of appearance, so different parts and surfaces of the body came to prominence. Hands and fingernails, for example,

DOI: 10.1057/9781137467485.0005

were important symbols of beauty and virtue. On the face, the most public of bodily surfaces, eyebrows were seen as barometers of character. The tweezers to maintain them were thus important items of toilette. As changing attitudes towards the smile rendered the teeth more visible, toothpicks and brushes were also essential pieces of kit. This chapter therefore adopts a different approach, exploring the broader importance of small instruments as progenitors of bodily transformation.

Chapter 4 explores the changing nature of spectacles in the eighteenth century. Steel-framed spectacles began to appear around the 1750s. Makers such as Benjamin Martin and James Ayscough utilised the springy strength of steel to transform the design of spectacles from their traditional armless pince-nez design, to a new form with side arms that used pressure to stay tightly adhered to the wearer's temples. Martin's new 'Martin's Margins' spectacles, introduced around 1760, could be highly polished to give a pleasing appearance, whilst other sorts of 'wig spectacles' were designed to help myopic macaronis attend society functions in comfort and safety.[35] As spectacles became more decorous, they also became more public. Whilst never becoming desirable items of fashion, they nonetheless shook off previous pejorative connections and became connected with learning, sagacity and the quest for knowledge through reading and 'seeing' the world.

The final coda chapter argues that surgical instruments, despite not being made or marketed for personal use, were nonetheless important in bodily transformation, bearing many similarities in design, manufacture and marketing with other instruments explored here. In the last quarter of the century, surgical instruments were increasingly shaped by steel. Its elasticity, durability and ability to carry a high polish rendered cast steel signally useful in surgical instrument manufacture. Like razor makers, cutlers and specialist surgical instrument makers engaged in a continual process of metallurgical experimentation and development. They were also expected to have some anatomical knowledge to cater for the exacting demands of their customers in the medical faculty. Changes in medical education, in particular the growth in anatomical dissection, drove demand amongst surgeons and surgical students for instruments. Makers appealed to 'gentlemen of the faculty' in their advertisements. But surgeons themselves also engaged in the development, refining, redesigning and patenting of instruments. In some cases, such as amputation, the changing form of instruments effected changes to surgical technique. But the patient's body was at the heart of change.

DOI: 10.1057/9781137467485.0005

Recognising the discomfort of surgery, and the collateral damage and pain caused by botched operations, surgeons called for instruments that utilised the properties of steel and were tailored to specific needs of medical procedures. Some manufacturers pitched their products, such as sharper lancets for bloodletting, on the basis that they reduced pain, thus attracting the approbation of the patient.

Overall, the book offers a new perspective on the important relationship between steel instruments and the purposeful management of the body during the Enlightenment. This was a period in which perhaps a greater range of goods was available for the body than ever before. Steel was an enabling material, one that changed both individual instruments and the practices and meanings associated with them. At the same time, cultural and religious shifts removed earlier taboos about interference with God's work. As new corporeal ideals were made and remade, people had both the motivation and the means to transform their own bodies.

By no means was steel the only material in this process; many other metals and other substances from rubber to ivory played important roles. But, both its physical properties and unique status as an 'enlightened metal' meant that steel was a key component in bodily refinement in eighteenth-century Britain.

Notes

1   David M. Turner and Alun Withey, 'Technologies of the Body: Polite Consumption and the Correction of Deformity in Eighteenth-Century England', *History*, 99:338 (2014), 781.

2   Advertisements, *World*, 18 May 1790.

3   Celina Fox, *The Arts of Industry in the Age of Enlightenment* (New Haven, CT: Yale University Press, 2010).

4   John Styles, 'Product Innovation in Early Modern London', *Past and Present*, 168:1 (2000), 124–169; M. Jones, 'Industrial Enlightenment in Practice: Visitors to the Soho Manufactory, 1765–1820', *Midland History*, 33:1 (2008), 68–96.

5   See Chris Evans and Goran Ryden, *Baltic Iron in the Atlantic World in the Eighteenth Century* (Leiden: Brill, 2007); Kenneth Barraclough, *Steelmaking before Bessemer Volume 2: Crucible Steel – the Growth of Technology* (London: IOM, 1984).

6   Parts of this discussion are taken from Alun Withey, 'Shaving and Masculinity in Enlightenment Britain', *Journal of Eighteenth-Century Studies*, 36:2 (2013), 225–247, and are used by permission.

DOI: 10.1057/9781137467485.0005

7   T. Craddock and J. Lang, 'Crucible Steel – Bright Steel', *Historical Metallurgy*, 38:1 (2004), 36.

8   Marcia Pointon, 'Jewellery in Eighteenth-Century England' in Maxine Berg and Helen Clifford (eds.), *Consumers and Luxury: Consumer Culture in Europe 1650–1850* (Manchester: Manchester University Press, 1999), 120. John Brewer and Roy Porter (eds.), *Consumption and the World of Goods* (London: Routledge, 1994); Jon Stobart, 'Selling (through) Politeness: Advertising Provincial Shops in Eighteenth-Century England', *Cultural and Social History*, 5:2 (2008), 309–28; Amanda Vickery, *Behind Closed Doors: At Home in Georgian England* (New Haven and London: Yale University Press, 2009).

9   See Styles, *The Dress of the People*, esp. 97–107.

10 Joan Evans, *A History of Jewellery 1100–1870* (London: Faber and Faber, 1970 edition), 161.

11 Michael Polanyi, *The Study of Man* (Chicago: University of Chicago Press, 1966 edition), 25.

12 Liliane Hilaire-Pérez and Christelle Rabier, 'Self Machinery? Steel Trusses and the Management of Ruptures in Eighteenth-Century Europe', *Technology and Culture*, 54:3 (2013), 460–502. See also Chris Evans and Alun Withey, 'An Enlightenment in Steel?: Innovation in the Steel Trades of Eighteenth-Century Britain', *Technology and Culture*, 53:3 (2012), 533–60. See also Bryan S. Turner, *The Body and Society*, second edition (London: Sage Publications, 1996), 20–1, 23; Adelheid Voskuhl, *Androids in the Enlightenment: Mechanics, Artisans and Cultures of the Self* (Chicago: University of Chicago Press, 2013), 1–2, 8, 10; for automatons, see also Simon Schaffer, 'Enlightened Automata' in William Clark, Jan Golinski and Simon Schaffer (eds.), *The Sciences in Enlightened Europe* (Chicago: University of Chicago Press, 1999), 126–65.

13 Perez and Rabier, 'Self Machinery', 462.

14 Lynn Sorge-English, Stays and Body Image in London: *The Staymaking Trade, 1680–1810* (London: Pickering and Chatto, 2011).

15 For example, Mark Jenner and Patrick Wallis (eds.), *Medicine and the Market in England and Its Colonies, 1450–1850* (London: Palgrave, 2007); for specific examples, see Roy Porter, *Quacks: Fakers and Charlatans in Medicine* (Stroud: Tempus, 2001); Colin Jones, 'The Great Chain of Buying: Medical Advertisement, the Bourgeois Public Sphere, and the Origins of the French Revolution', *The American Historical Journal*, 101:1 (1996), 13–40. For other examples of the relationship between bodies and mechanisms in various contexts, see George Vigarello, *Histoire Du Corps Volume 1* (Paris: Seuil, 2006); Colin Jones, *The Smile Revolution in Eighteenth-Century Paris* (Oxford: OUP, 2014); Carsten Timmerman and Julie Anderson (eds.), *Devices and Designs: Medical Technologies in Historical Perspective* (London: Palgrave Macmillan, 2006).

16 See Nicholas Andry, *Orthopædia: or, the Art of Correcting and Preventing Deformities in Children* (London: 1743); for personal grooming, see David M.

DOI: 10.1057/9781137467485.0005

Turner, 'The Body Beautiful' in Carole Reeves (ed.), *A Cultural History of the Human Body in the Enlightenment* (London: Bloomsbury, 2010), 130; Lynn Festa, 'Cosmetic Differences: The Changing Faces of England and France', *Studies in Eighteenth-Century Culture* 34 (2005), 25–54; Richard Corson, *Fashions in Makeup* (London: Peter Owen, 1972).

17   Liliane Perez, 'Technology Curiosity and Utility in France and England in the Eighteenth Century' in Bernadette Bensaude-Vincent and Christine Blondel (eds.), *Science and Spectacle in the European Enlightenment* (Farnham: Ashgate, 2008), 29.

18   Turner and Withey, 'Technologies', 786, 795.

19   Lawrence Klein, 'Politeness and the Interpretation of the British Eighteenth Century', *Historical Journal*, 45:4 (2002), 869–70.

20   John Brewer, *The Pleasures of the Imagination: English Culture in the Eighteenth Century* (London: Harper Collins, 1997), 102.

21   Klein, 'Politeness', 869–98.

22   Ibid., 870; Susan Fitzmaurice, 'Changes in the Meanings of *Politeness* in Eighteenth Century England' in Jonathan Culpeper and Daniel Z. Kadar (eds.), *Historical (Im)Politeness* (Bern: Peter Lang, 2010), 87–8.

23   Fitzmaurice, 'Changes', 89.

24   See John Tosh, *Masculinities in Britain Since 1800* (London: Routledge, 1991); Tim Hitchcock and Michèle Cohen, *English Masculinities, 1660–1800* (London: Routledge, 1999); Philip Carter, *Men and the Emergence of Polite Society, 1660–1800* (London: Routledge, 2000); Michèle Cohen, '"Manners" Make the Man: Politeness, Chivalry and the Construction of Masculinity, 1750–1830', *Journal of British Studies*, 44:2 (April 2005), 312–329; Styles, *The Dress of the People*; Pointon, *Brilliant Effects*.

25   Karen Harvey, *The Little Republic: Masculinity and Domestic Authority in Eighteenth-Century Britain* (Oxford: OUP, 2014); Vickery, *Behind Closed Doors*.

26   Helen Berry, 'Polite Consumption: Shopping in Eighteenth-Century England', *Transactions of the Royal Historical Society*, 12 (2002), 375–94; Maxine Berg and Helen Clifford, 'Selling Consumption in the Eighteenth Century: Advertising and the Trade Card in Britain and France', *Cultural and Social History*, 4:2 (2007), 145–70; Stobart, 'Selling (Through) Politeness'.

27   Richard Sennett, *The Fall of Public Man* (London: Faber and Faber, 1976), 64–5.

28   Samuel Tissot, *Three Essays: First on the Disorders of People of Fashion, Translated by Francis Bacon* (Dublin: printed for James Williams, 1772), 37–9.

29   David Hartley, *Observations on Man, His Frame, His Duty, and His Expectations. In Two Parts, Volume 1* (London: Printed by S. Richardson, 1749), iv.

30   Anon, 'An Essay on the Nature, Marks and Principles of Politeness', *The Universal Magazine*, December 1775, 308.

31   Ibid.

32 Turner and Withey, 'Technologies', 788.

33 Carter, *Men*, 29; G.J. Barker Benfield, *The Culture of Sensibility: Sex and Society in Eighteenth-Century Britain* (Chicago: University of Chicago Press, 1992), xvii-xx.

34 Ibid.

35 J. William Rosenthal, *Spectacles and Other Vision Aids: A History and Guide to Collecting* (San Francisco: Normal Books, 1996), 41, 99–105.

DOI: 10.1057/9781137467485.0005

1

Shaping the Body: The Politics of Posture

Abstract: *Bodies that were deformed, crooked or twisted were stigmatised in the eighteenth century. Apart from inviting ridicule, a body that was not 'ideal' could inhibit social prospects. Changing ideas about the body laid new emphasis on 'correcting' the vagaries of nature and restoring it to a 'natural' form. The elastic properties of cast steel rendered it a useful component in corrective devices. Makers who styled themselves as body specialists, rather than medical practitioners increasingly used it. Such devices ranged from trusses to treat hernias, to machines to correct posture and promote upright stance and 'straightness'. The treatment of children and adolescents was seen as especially important for their future prospects. Devices were often extremely uncomfortable for users, as well as unsightly. Charting contemporary debates about the nature of ideal body shapes, Withey explores the eighteenth-century paradox of using unnatural means to achieve a natural shape.*

Withey, Alun. *Technology, Self-Fashioning and Politeness in Eighteenth-Century Britain: Refined Bodies.* Basingstoke: Palgrave Macmillan, 2016. DOI: 10.1057/9781137467485.0006.

> *There are few diseases which afflict the Human Body, attended with greater disadvantages, than those produced by Distortion. It gives not only an unpleasing appearance, but innumerable complaints generally follow.*[1]

As the above statement from an 1800 advertisement suggests, deformed, 'crooked' or twisted bodies could seriously impede the social ambitions of their owners. Deformity was not merely physically discomforting. It also portended serious social consequences, not least in the mockery often meted out to the disabled. Many slang terms and insults were levelled at those whose physical appearance was enough to draw glances. According to James Caulfield, author of a 1793 dictionary of slang, 'these unhappy people afford great scope for vulgar railleries'.[2] Imagining an encounter with a 'crooked or hump-back'd person', Caulfield detailed the sorts of insults that could be let loose. 'Did you come straight from home? If so, you have got confoundedly bent by the way', went one example. 'Don't abuse the gen'man', adds a bystander, 'he has been grossly insulted already ... don't you see his back's up?'[3] Besides hurt feelings, there were more material concerns. Physical impairment could hinder women's prospects of marriage, not to mention her hopes of romantic ardour. Some eighteenth-century singles' advertisements even stipulated that the prospective match should be, as one put it, 'without any deformity in her person'.[4] Many afflictions were begotten by the conditions of daily life or simply through sloppy habits of posture. Occupations and pastimes requiring close work, bending and stooping could injure the shape of the human form, amongst them 'holding down the head; putting out the chin; stooping in the shoulders; bending too much forwards and thrusting out the belly'.[5]

The period after 1750 was transformative in terms of attitudes towards bodily ideals, and the extent to which intervention was permissible and desirable. Work by Georges Vigarello, for example, has explored the early modern change from a culture in which children's bodies required shaping, to one in which they were left to nature.[6] However, the use of stays by women to enhance the visual aesthetic of their bodies increased and, by the 1760s, was integral to female bodily transformation.[7] David Turner and others highlight the multifarious meanings of impairment in the long eighteenth century, and the various ways in which missing

DOI: 10.1057/9781137467485.0006

limbs, bent backs and twisted bodies were understood and articulated.[8] As Turner notes, the use of terms like 'lame', 'crippled' and 'deformed' reflected broader connections of impaired bodies with 'monstrosity'.[9] The lexical focus lay firmly upon bent, misshapen and otherwise highly visible deformity. There were, however, also lesser degrees of impairment. Whilst no formal category existed in contemporary terminology, many bodies were irregular, rather than deformed. In his 1754 essay on disability, the MP William Hay, himself a sufferer of a spinal condition and restricted height, differentiated between degrees of disability. For Hay, spinal conditions were fundamentally different to, say, deafness or blindness, each of which resulted from different causes and had its own unique consequences.[10] Hay also defined deformity as 'visible to every eye', suggesting a condition that was manifest and hard to disguise.[11] A combination of poor diet, hard working conditions and lack of adequate medical intervention into congenital diseases and defects in childhood further served to mark the body. In this sense much of the population probably inhabited bodies that did not conform to any sense of an ideal.

Many less acute conditions could be disguised to imitate a natural form. Many new devices aimed at subtle correction rather than root and branch reshaping. Indeed, as will be discussed, many bodily technologies were sold upon their ability to be dicreet, allowing the wearer to be indistinguishable from nature. Alongside correction was the second important issue of 'improvement'. Whilst the visibly disabled and deformed perhaps formed the primary market for corrective devices, advertisers tapped popular fears about body image and addressed those seeking cosmetic 'improvement' of their own bodies, as well as preventative measures. Into this latter category fell parents keen to train their children's posture.

In the early eighteenth century, the signs and symptoms of an impaired body might represent vagaries of nature, for which individuals held responsibility to correct.[12] For others correctional devices went *against* nature, and debates raged about the health risks involved in reshaping the body, as well as the moral consequences of bodily reshaping through vanity. Some viewed stays, collars and other corrective devices as deceitful, creating the mere phantasm of a natural body that evaporated once the person disrobed, revealing their true state. Constant tension existed between the quasi-medical need to prevent or correct deformity, and the social and cultural background of improvement. Nevertheless, by the second half of the eighteenth century, cultural shifts meant that the use

DOI: 10.1057/9781137467485.0006

of artificial means to restore or improve the body was no longer frowned upon. In the midst of a growing consumer market, bodily technologies could reflect fashion as well as function.[13]

Metallurgical innovation, particularly in steel, afforded new possibilities for bodily shaping. Recent work has focussed on the part played by steel in the construction, marketing and consumption of hernia trusses in the eighteenth century.[14] Far less attention, however, has been paid to its broader impact upon devices for altering posture. Steel's physical properties made it perfect for devices where support or compression was needed. But unlike the advertising of other products in this book, such as razors, the use of cast steel was not, of itself, a marketing tool for postural devices. Although makers sometimes claimed it as a superior alternative to other types of elastic materials, neither the 'scientific' nor 'philosophical' credentials were emphasised. Likewise, while makers of razors and spectacles could highlight the aesthetic appeal of their products, the emphasis in the marketing of postural devices was more often upon concealment and subtlety.

Importantly too, while medical professionals originally claimed dominion over the prescription, fitting and application of correctional devices, specialist makers began to circumvent practitioners, marketing their products directly at lay consumers. As this occurred, people took greater control over shaping their bodies. Debates raged about the efficacy of home-fitted devices and the potential dangers of untrained users. Nevertheless, the market for correctional devices, from stays and collars to trusses and neck swings, expanded markedly.

The 'natural' body

Identifying bodily ideals in the eighteenth century is difficult, not least because of the absence of firm evidence that any such ideal was ever codified. Were changing ideas about the body consistent across all of society, or were different mechanisms at play below elite levels? It seems probable, though by no means certain, that the primary audience and market for postural devices were middling orders and elites, who had the disposable income, access to shops and advertising and also, perhaps more importantly, the impetus to fashion their bodies to suit normative rituals of politeness. Nonetheless it is unsafe to assume that the lower orders had no desire to participate in the improvement of their bodies.

DOI: 10.1057/9781137467485.0006

Deformity was not socially delimited, and the desire to escape ridicule was surely a powerful incentive even if there was less demand to conform to expectations of gentility.

In shaping or 'correcting' their bodies, contemporaries were not necessarily explicit in seeking an ideal. Neither medical nor lay texts about the body, or indeed the marketing of devices, explicitly defined one. And yet people seemingly knew what they wanted their bodies to look like and, indeed, what they wished to avoid. If there was a new bodily ideal, then, it seems to have been articulated through growing cultural emphasis upon corporeal aesthetics, including neatness, symmetry and elegance. Once, the owners of impaired bodies were largely powerless, and indeed actively discouraged, to interfere with God's intentions. But gradual change after 1750, encompassing to some degree the secularisation of the body, accompanied by the burgeoning market and availability of devices, meant that authority and impetus for change lay with the individual.

At the heart of the rhetoric of correction was the somewhat chimeric concept of a 'natural' body shape.[15] For some, 'natural' represented the body in a state of perfect nature, in other words unadulterated and unadorned. This implied that unaltered bodies were essentially most desirable. In the 1743 *Orthopaedia*, Nicholas Andry repeatedly referred to 'natural' bodily features as being those unaltered by artifice.[16] The postural specialist and truss maker Timothy Sheldrake Junior suggested that 'any attempt to improve the natural shape by art' was doomed to failure. Women who had never worn stays, he argued, were better shaped, and deformity was never seen amongst women 'nearest a state of nature', implying those who shunned devices and simply left their bodies alone.[17] But even a cursory glance around the town square would have confirmed that nature was all too capable of begetting malformed specimens. For such people, with the means and desire to do so, the only recourse was to a device to correct or conceal their true shape. The obvious paradox, then, was that people sought a variety of devices to mimic a 'natural' body shape; a concept that was itself predicated on the notion of being unadulterated and unaltered.

Since the makers of correctional devices often claimed to be *restoring* a body to its natural form, what exactly were the keystones of a 'natural body'? Proportion was important. Andry codified the proportional measurements of various bodily parts, and their relationship to others. Whilst acknowledging that 'nature varies very much', and that body shapes and forms were diverse, he argued that nature automatically compensated

DOI: 10.1057/9781137467485.0006

for variations in one part by maintaining the balance of proportion in another.[18] Thus, a body with long legs would be compensated with a shorter upper body. To 'remedy' nature made little sense to Andry since clumsy human attempts would merely upset the 'exact symmetry' of the body, rendering the deformity the more monstrous.[19]

Bodily appearance was supplemented by genteel deportment, behaviours and manners. Conduct literature suggested that an 'easy posture' was needed to create an elegant and harmonious impression upon viewers. In 1762, the *Polite Academy* instructed youngsters on correct form, including how to hold posture to achieve a 'genteel figure'.[20] Adolescents should stand 'free and easy' with head upright since 'to be stiff is almost as bad as to stoop'. To 'poke the head forward' deformed the back, making a body 'appear vulgar and ill-shaped'.[21] Dance was vaunted as a useful expedient for good posture, and dancing masters were acknowledged experts on deportment, and educators in the principles of polite behaviour. Medical practitioners and philosophers alike recommended dancing, and other forms of vigorous exercise, as useful and healthy expressions of 'natural tendencies'.[22] By the end of the century, dancing masters promoted stiff, erect postures as the modern trend.[23] John Weaver's *Anatomical and Mechanical Lectures upon Dancing* sought to explain bodily proportions 'according to the absolute rules of nature; that we may have before us a standard by which to measure the rest'.[24] Especially important, both for aesthetic and functional reasons, was straightness. The spine was the mainstay of the body, likened both to 'a kind of trunk' and the keel of a ship. When straight, well set and proportioned, it presented a 'handsome body'. When 'crooked and ill turn'd', the body was deformed.[25] Straightness, then, bespoke gentility, while a deformed body was vulgar.

Debates surrounded gendered ideals of bodily form. Some saw the uniform curve, rather than the ramrod-straight spine, as beauty's true marker. In his 1753 *Analysis of Beauty*, William Hogarth argued that 'the common notion that a person should be as straight as an arrow' was erroneous.[26] Hogarth argued that the curve, the so-called 'line of beauty', found across nature, was more elegant, composed and pleasing to behold.[27] For women in particular, attention was focussed upon achieving or augmenting curves. Lynne Sorge-English has noted that between 1775 and 1785, women's stays were purposefully designed to meet this 'serpentine' line, emphasising waist and breasts.[28] The physician Erasmus Darwin opined that a 'stiff erect attitude … does not contribute to the grace of person, but rather militates against it'.[29] Andry identified several differences between male and female bodies, arguing that women bore slenderer waists and higher bellies than men, as well as thicker legs – 'yet this is by no means a perfection'.[30] The situation is further complicated by a lack of evidence for the supposed ideals for men's bodies. Whilst attaining a curve may have been desirable for women, men's bodies were expected, at least before the 1770s, to personify strength and tone. Some authors stuck to vague generalisations. Weaver noted simply that the symmetry and proportion of men and women differed, with women being smaller than men and with 'remarkably narrower shoulders'.[31] Ideals of body shape were therefore chimeric and subject to debate and change. Even if an ideal remained elusive, however, there was at least a general agreement that shape, form, neatness and harmony were important in conveying the underlying character of the individual.

Truss makers and postural specialists

The marketplace for correctional devices was diverse, and the range of makers no less so. In the seventeenth century, the supposed boundaries even between medical trades were liminal. Surgeons often ran apothecary shops; barbers and barber-surgeons were largely interchangeable; and medical practice was ubiquitous across the country. As the marketplace expanded, there were moves towards greater specialisation in medical trades, and especially artisans making and selling their own products.[32] Truss and stay makers were relatively early arrivals, manufacturing various postural and correctional devices. As we will see in other chapters, specialists from oculists to razor makers occupied a shadowy position in the medical market, claiming to cure, but not necessarily positioning themselves as medical professionals.

The place of truss and stay makers was ambiguous. Some makers of correctional devices deliberately positioned themselves as suppliers to the medical faculty, without necessarily implying that they, themselves, were practitioners. The 1734 trade card of truss maker and stay maker James Lane addressed 'Physitians and Surgeons', while 'Mr Dowling of Duke's court' in 1790 spoke to 'Gentlemen of the Faculty'.[33] Truss maker Robert Brand wrote *Chirurgical Essays*, endorsed by the king's surgeon and medical luminary, John Hunter.[34] Most, however, addressed individual sufferers and the wider public, clothing their appeals in the genteel language of

DOI: 10.1057/9781137467485.0006

polite commerce. Some had 'perfect confidence' in popular approbation of their products, while others 'begged leave to inform the public' of their latest invention. One even felt that he would be neglecting his own duty if he did not promote his wares to the public.[35] Occasional tensions arose between makers of corrective devices and medical practitioners, who saw the former as dangerous, untrained meddlers. Brand advocated a symbiotic relationship wherein truss makers designed premium products using their specialist knowledge, but left it to surgeons to apply.[36] In response, some makers acknowledged the danger of quackery – mostly in others – and positioned themselves instead as body specialists who supported medicine.

Years of practical experience and successful treatment were one way of claiming authority, while emphasising training by an eminent surgeon or receiving the approbation of the faculty were others.[37] Some even based their pitch on rescuing the victims of empirics. 'It is well known by the faculty', stated London truss maker J. Eddy in 1797, 'that every pretence by Quacks, with their nostrums and external applications' to cure patients was 'a great deception on the afflicted'.[38] Only a specialist, with the requisite knowledge of the body and the nature of deformities, should be sought. Eddy claimed regular consultations with those whom quacks had failed.[39] A little puffery also went a long way. Aware of an apparent reputation for being the 'herald of his own fame', Eddy (now stressing his M.S.D. qualification) assured the public that his 'self-applause [was] equalled only by his abuse of all who pretend to cure ruptures, except himself'.[40]

Makers enthusiastically stressed their own artisanal skills and use of new, innovative materials. Diversification reassured the public of the broad range of knowledge possessed by individual makers. Being knowledgeable and skilful across different techniques, and aware of the theory and philosophy behind his trade, elevated the maker above the mere dilettante or 'mechanick'. As with other areas of specialist manufacture, postural device makers experimented with new materials, and collaborated with other trades, to improve their products. Truss making, for example, might involve input from purse makers, cutlers, locksmiths and tailors.[41] Like razor makers, some used their metallurgical and engineering expertise to diversify into other trades. Spinal specialist Sheldrake, for example, applied for a patent for new types of wheels to facilitate movement in engines and machinery.[42] By 1800, the importance of makers of corrective devices as innovators was

DOI: 10.1057/9781137467485.0006

acknowledged by medical professionals, who sought to bring them back to the fold. James Lucas, a surgeon of the Leeds Infirmary, lauded the successes made when 'regularly trained practitioners' contrived specialist machines, and made clear both the usage and means of manufacture. Referring to the utility of 'iron and steel made elastic by rolling', Lucas suggested that 'mechanical surgeons' should draw on expertise outside medicine, including 'the ingenious contrivances of Messrs. Bolton and Watt' – i.e. the Birmingham industrialists and Lunar Club founders Matthew Boulton and James Watt.[43]

Steel devices competed with other types of contrivances, the makers of which claimed superiority. Elastic devices like trusses and supports for 'contracted limbs', incorporating fabric, springs and, later, rubber, were generally cheaper and, according to their makers, as good, if not better, than steel models. Indeed, some stressed the *lack* of steel in manufacture as an advertising device. In 1773, for example, Dr De Malon promised to cure 'rickety and crooked children without the Application of any Steel Instrument'.[44] Truss maker Robert Brand promoted his 'True Elastic Truss' as a superior alternative to those 'dignified with the names of Steel and Spring trusses', while Eddy's elastic products were 'different from any yet made'.[45] But the springy, tensile strength of steel was extremely well suited to shaping the body, as well as becoming a desirable material in its own right, the use of which could be emphasised by makers. Steel was therefore a vector through which new ideals of bodily form and appearance could be achieved and maintained.

Children

Disability or deformity in children was particularly troublesome for parents. It put added strain upon family life due to the extra care needed and financial implications of paying for treatment.[46] There were also social and financial implications of bodily appearance and deportment for the child itself in later life, including employment and marriage prospects. It is clear, though, that many parents, and by no means just elites, invested much in their disabled offspring and sought to nurture and provide for them.[47] Before the 1750s, views of children's bodies still reflected sixteenth-century ideas of malleability and tenderness. The sixteenth-century practitioner Felix Wurtz compared children's bodies 'to a young and tender root or twigg of a Tree, which in the souch is

DOI: 10.1057/9781137467485.0006

not so grosse as an old root or branch'.[48] Eighteenth-century medical treatises still advocated firmness but tenderness with infants' bodies so as not to 'bring crookedness upon them', and advised parents on swaddling infants correctly to promote a straight body.[49] Andry's *Orthopaedia*, the first to deal explicitly with the correction of childhood deformity, strongly advocated not binding the body too tightly for fear of distorting the bone structure.[50] Clothing was important in this process. Stays (or corsets), reinforced with bone, were available for children as young as two, and designed to force a child's body into the desired, erect form, and promote an upright gait.[51] Meares of Ludgate Hill even advertised his trusses as being 'so easy that a child of a Month old may daily use them without pain'.[52]

Some quasi-medical practitioners designed and built their own 'machines' specifically for relieving children. In 1768, D. Merande, lately arrived from Paris, regarded childhood deformity as 'the most alarming kind' and his 'newly invented machine for the cure of preternatural curvatures of the spine' supposedly palliated symptoms ranging from hiccoughs to loss of sleep.[53] With some childhood diseases like rickets or club foot, new types of devices sought to complement medical remedies. Early eighteenth-century practitioners were as likely to treat 'rickety children' with products such as 'chymical drops' or 'worm plaisters' as to try and rectify the child's gait.[54] Medical treatises, quoting luminaries like Herman Boerhaave, suggested health regimens including light diet, warm clothes, cold baths and frequent coach rides for fresh air and to promote friction against the skin.[55] No mention was made of specific technologies to straighten the legs. In 1743, Andry referred to 'several machines proposed for exercising rickety children', but also advocated medicines and good diet.[56] By the late eighteenth century, however, the use of corrective technologies was fairly widespread. Treatments for club foot ranged from applying tight bindings and leg irons to placing the feet in specially designed compression boxes, or tying metal plates to the feet and placing them in special shoes containing a 'ratchet wheel', which, by turns, wrenched toes back to their 'natural shape'.[57]

But children, especially infants, lay at the heart of debates about bodily shaping. Whilst postural device makers lauded the straight body, others railed against the interference of overzealous parents. W.B., a Paris surgeon, bemoaned the 'ridiculous experiments' visited upon children by mothers and nurses, causing the plague of 'deformities and distempers' including 'hump backs, crooked bodies [and] shoulders awry',

DOI: 10.1057/9781137467485.0006

particularly prevalent in London.[58] The letters of John Hill's imagined gentlewoman denounced mothers who used 'steel armour' on children, inevitably bestowing upon them a lifetime of deformity. Obsessed with their own child's appearance, they 'pity the poor who cannot take care of their children's shapes'.[59] Sheldrake agreed, arguing that childhood deformities were easily preventable by simply 'follow[ing] the dictates of reason' instead of applying devices and techniques about which parents understood little.[60]

For older children, all manner of devices promised preparation for polite society by addressing posture and gait. Common were devices concealed within clothing, which made it difficult, indeed painful, for the adolescent to slouch. The London truss maker John Sleath advertised many such products in the 1790s, including steel 'backs and collars' and 'monitors', all of which involved rigid metal plating.[61] Backs and monitors kept the spine straight, while collars thrust the chin upwards. These could be extremely painful, especially for children. One letter in an imagined series from a clergyman to his son alludes to the 'conduct of the matron, who, to prevent her daughter from dropping her chin into her bosom, threw it up into the air by the aid of a steel collar'.[62] In an apparent reference to the entertainers and ladies of ill repute for which the area was infamous, another complained that the steel collar 'pinioned [his] shoulders further back than the people that go up Holborn Hill'.[63]

Good posture in adolescents, particularly girls, was considered extremely important. The truss maker Eddy offered 'leg irons for crooked and weak-kneed children' together with 'steel backs and collars for young ladies on an improved principle' and 'spinal stays for rickety children'.[64] By 1811, some makers even supplied schools and boarding houses. Advertisements for Talmage's 'Fashionable Corset Warehouse' in Portman Square, London, noted that 'Boarding schools were accommodated on reasonable terms'.[65]

Correction and concealment

Given the lack of evidence for usage, accessing the motivations of consumers, the popularity or the experience of wearing devices is problematic. Complaints against the practice, however, suggest that wearing them was common enough to attract censure. In 1753, for example, a correspondent to the Newcastle *General Magazine* complained about the 'artificial deformity' adopted by young ladies who deliberately altered their shape for fashion.[66] For the correspondent, the transformed body was itself rendered artificial, modelled by fashion rather than taste. As evidence for the deception, a list of culpable (and in some cases fantastic) tools was implicated, including 'head moulders, face-squeezers, nose-parers, ear-stretchers, eye-painters, lip-borers, tooth-stainers, breast-cutters, foot-swathers, &c, &c'.[67]

But despite the various moral objections raised, the expanding numbers of artisans, and products, suggest a practice that was increasing after 1750.[68] By the 1780s, Sheldrake was noting that the present age was 'the time of life that a fine shape is of the utmost consequence to the fair sex'. For those unfortunate to suffer any sort of deformity, 'the most desirable object is to conceal the defect, and preserve the appearance of an elegant shape, when we cannot obtain the reality'.[69] The privileging of concealment and appearance over symptomatic relief here is noteworthy. For women in particular, creating the 'elegant shape' was of the utmost importance. Various products and devices targeted 'distortions' of the human body. One Jones of Tottenham Court Road, London, advertised 'Spinal Stays and Machines', as well as puffing his successes in 'restoring children from a distorted to a *proper* (my emphasis) shape'.[70] Jones, by his own admission, was no medical practitioner but a stay maker, albeit one with the approbation of 'several medical gentlemen'.[71] Here again, the 'proper shape' highlights shifting views about the impermanence of deformity. Some authors indeed stressed that disability and deformity were neither the sufferer's fault nor a reflection on their inner virtue, although the 'otherness' of the impaired body remained deeply entrenched.[72]

Just as for children, many devices were available for adult bodies. Widow Johnson of Little Britain, London, advertised her stock of collars, steel bodices and 'polish'd steel backs', alongside other instruments for the 'lame, weak or crooked'.[73] Such devices were often cruciform in shape, attached by leather straps around the arms and upper body, and acted to force the back and shoulders into the desired position. Sheldrake's 'Patent Elastic Back Collar' relied upon the 'laws of action and reaction' via a spring 'to draw the parts more forcibly into their natural situation'.[74]

Amongst the means employed to promote adult straightness, spinal devices were prominent. Whilst spinal curvature diminished the general *mien* of the body, it also invited illness by compressing the internal organs and causing vertebrae to fuse. Some professionals stressed that

DOI: 10.1057/9781137467485.0006

curvatures could result as much from 'improper attitudes', that is, a stubborn predilection for standing incorrectly, than from acute illness or congenital deformity. The impacts of hard labour or life stage were also regarded as factors.[75] Nonetheless, recovery of health was not necessarily privileged over the restoration of form. Various devices aimed at correcting spinal deformity by using the patient's body weight to stretch their spine and separate the vertebrae.[76] So-called 'neck swings' were the first incarnation of these machines, and were available by 1740. Under a physician's direction, the patient was lifted off the ground, suspended by the neck in a fixed apparatus. Whilst the neck swing was not something bought by individuals for domestic use, its primary function was to improve and correct, rather than treat. Various spinal specialists looked to improve the design, arguing that portable machines might be more beneficial. Sheldrake claimed to improve upon an earlier design of Monsieur Vacher in designing a system of metal springs, plates and cap, to extend the spine.[77] Another supposed innovation, the 'screw chair', was little more than a neck swing attached to the back of a chair, into which the patient was interned for up to ten hours per day.[78]

Other devices sought to improve posture through exercise. Movement and exercise were established features of humoral medicine as part of the emphasis upon health regimens. Eighteenth-century medical authors also viewed exercise favourably, the more vigorous the better.[79] This chimed with mechanistic views of the body as a living machine. Also important, however, was an emphasis upon the importance of the 'non-naturals' (air, diet, drink, motion and rest, sleep, evacuation and the passions), the regulation of which determined personal health, and which were a feature of medical advice books in the late eighteenth century.[80] It might be speculated that the emphasis upon movement in the treatment of bodily impairment fitted this framework well, in restoring static, and therefore effectively resting, bodies to motion.

As in a machine, 'a trifling irregularity or impediment in one part, may disturb the equal motion of the whole'.[81] But exercise was also promoted for reinvigorating tired or wasted limbs, and especially treating gout. John Cheshire's 1747 *Gouty Man's Companion* alluded to 'contrivances for the exercise of the gouty limb'.[82] Something of the appearance and attitude of gout sufferers who adopted various types of orthopaedic apparatus can be seen in Paul Sandby's satirical depiction in Figure 1. Posture was central to the physician Francis Fuller's arguments about the healthy body, with an entire volume dedicated to exercise. 'An erect

DOI: 10.1057/9781137467485.0006

position', Fuller argued, was 'essential to the well being of the body of Man'.[83] Without it, a person was reduced to a 'mere bed-ridden creature'.[84] Devices were available to bend and stretch bodies. Motion, especially such as swinging, which caused the organs to move, was especially recommended.[85] In 1779, Abraham Buzaglo lodged a patent application 'for machines &c for Gymnastick exercises'.[86] 'Dumb bells for opening the chest and exercising the body' were available from various makers, which sought to streamline, correct and also warm the body.[87] Held in both hands, they were swung alternately, in the manner of real bells, using gravity, and the weight, to 'cure' the distortion.[88] In at least one case, however, over-vigorous use of dumb bells was blamed for causing an inguinal hernia.[89] Others sought to improve or aid bodily posture *while* a person was involved in exercise. One such was the 'elastic saddle', which made riding more comfortable for corpulent, gouty persons – precisely those for whom exercise would be most beneficial.[90]

FIGURE 1   *Three men wearing orthopaedic apparatus, 1783, by Paul Sandby*
Sources: Image courtesy of Wellcome Images.

DOI: 10.1057/9781137467485.0006

A key selling point of correctional devices was their ability to be indistinguishable from nature, or invisible to the most intimate companion. Unlike other items discussed in this book, such as spectacles, and even razors, which could have an aesthetic appeal beyond their function, correctional devices were neither fashionable nor decorous. Occasionally some makers added ornamentation to try to render them more appealing. William Palin's 'steel backs' for ladies, for example, were 'cover'd with Morroco', while the widow of Samuel Johnson sold steel backs that were 'polish'd'.[91] They were never intended, however, for public display. People did not wish to show (or perhaps even acknowledge) their reliance on devices to achieve their shape, even to close friends or family. Rather it was the illusion of a 'natural' body that mattered. Claims such as those made by Mr Parsons, stay maker of Covent Garden, that his products were imperceptible even to 'the most intimate acquaintance' reinforce the importance of 'invisibility'.[92] Palin's steel backs were 'calculated entirely to the shape of the wearer', allowing them to be completely disguised beneath clothing.[93]

Some makers and retailers also stressed discretion in fitting. Ladies wary of an intimate examination were reassured that the wife of the retailer would attend to them, allowing modesty to be preserved.[94] For those for whom the prospect of physical examination was unconscionable, some even offered a postal service whereby a simple note of the dimensions of the rupture would result in a bespoke device being sent. Such practices raise the paradoxical issue of shame. People looked to artificially correct or conceal some bodily aberration of which they were self-conscious whilst, at the same time, avoiding being seen in devices that might further draw attention to themselves. Mail order could be viewed as a simple means of avoiding a potentially costly and inconvenient visit to a London specialist. But, it also allowed self-conscious sufferers to circumvent a potentially embarrassing examination. This was, therefore, a market that was both public and clandestine.

Little has been said about the experience of wearing postural devices, but much evidence points to their being cumbersome, uncomfortable and even painful. Users of neck swings, for example, were particularly vulnerable, not just to physical discomfort but also the loss of dignity caused by being dangled above the ground by the neck. Parisian visitor D. Merande suggested that the initial delight experienced by children at being 'balanced in the air' soon gave way to weariness and muscular fatigue.[95] Users suffered painful effects from the tight confinement,

DOI: 10.1057/9781137467485.0006

including complete limb numbness. Merande also noted the pain caused by pressure upon extruding vertebrae, and the general discomfort of any forms of bodily compression.[96] Many devices were based on the principle of restriction of movement, or the painful forcing of limbs into the desired position.

The very materials from which devices were manufactured, especially metals, could render them uncomfortable, causing problems for advertisers. These ran from the sheer weight of steel machines to the besmirching of clothing including 'iron moulding on the most delicate muslin frocks'.[97] J. Sleath reassured ladies that his steel backs and collars 'of entire steel' were 'peculiarly light, neat and durable'.[98] Others caused friction, resulting in skin burns or blisters. The unfortunate Reverend Joseph Greene, forced by an accident to wear an 'elastick steel truss', described how the constant wearing of this 'uncouth bandage... rather bruises ye Contiguous parts; and is, at times, very troublesome to me'.[99] Writing in 1780, the surgeon Henry Manning commented on popular devices like steel stays, neck swings and screw chairs, which, he argued, were of little practical help. Indeed, according to Manning, patients frequently became unhealthy and died in an exhausted state, or were forced to live out a miserable existence confined to chair or bed. Deprived of the power of locomotion they were, in his view, useless to themselves and others![100]

A nineteenth-century history of the treatment of spinal deformity by Henry Heather Bigg gives a useful account of the experience of wearing postural devices. Bigg quoted an unnamed English woman who was treated for over twenty years by one Mr Chessher of Hinckley in Leicestershire a century earlier, 'to whom flocked the deformed of all classes from all parts of England'.[101] (Interesting to note here is the reference to 'all classes', suggesting that bodily transformation was not limited to the upper echelons of society.) The woman first consulted Chessher at age sixteen when she was, by her own admission, of 'good bodily health'. Her first treatment was to 'wear his steel collar, which conveyed the weight of the head upon the hips'. She also regular used the neck swing, and described what must have been an arduous experience. 'I remained suspended in a neck swing, which is merely a tackle and pulley fixed to the ceiling of the room; the pulley is hooked to the head-piece of the collar, and the whole person raised so that the toes only touch the ground'. In this position, she spent much of the day. Remarkably, after two decades of treatment, it was reported that her spine had actually

DOI: 10.1057/9781137467485.0006

decreased by six inches.[102] This was doubtless an extreme, and some makers were keen to stress the unobtrusive nature of their apparatus; one suggested that, whilst wearing his machine, men could walk around and employ themselves as they pleased, while young ladies were not hindered from playing the harpsichord.[103] In such cases, people could be reassured that the impact upon their daily lives, and particularly upon their participation in polite social rituals, would be minimised.

Conclusion

Cast steel's physical properties afforded new possibilities for shaping bodies in the second half of the eighteenth century. For devices reliant upon sprung tension, it was ideal. Whilst support devices such as trusses had been available for centuries, the period after 1750 saw new devices emerge for correcting, curing or improving the body, along with numbers of specialists making and applying them. This occurred at the same time that people were apparently more willing to purchase devices that they could use upon themselves, rather than rely upon the services of a medical practitioner.

Whether the impetus to shape the body, or the impact of steel came first is unclear. Did manufacturers exploit a cultural change that was already in existence, or was the availability and perhaps the desirability of new devices the driver of change? The former seems more likely. Whilst trusses, collars, neck swings and so on can be located within the broader nexus of domestic consumption, they do not follow patterns of desirability, either in the materials used or in the devices themselves. What mattered instead was the end result of a straight body, if possible achieved without the device even being visible. As such, postural devices followed a different path to others discussed in this book.

Some other issues must be raised. Firstly, the obvious discomfort reported by wearers, together with the issue of the concealment of devices, questions their true utility. Did people buy devices simply because they were available, and appeared to offer a shortcut to bodily transformation? Were they perfect examples of Adam Smith's 'frivolous trinkets', which said more about the acquisitiveness of the individual than their desire to truly change their form? It could be argued, and indeed was argued by dancing masters in particular, that good posture began with training, not superfluous machines. Many objections were

DOI: 10.1057/9781137467485.0006

raised to the use of devices and, in particular, to those who used them as their first method of recourse, merely exacerbating their own problems by applying (or misapplying) them. Implicit amongst these objections was the assumption that sufferers were merely substituting one form of deformity for another.

The issue is further complicated by the lack of firm evidence for consumption. That a wide range of devices was available is clear. Advertisers tapped popular concerns about body image, stressing the importance of posture, the social drawbacks of deformity and, in particu-lar, the future prospects of young people. The medical debates about the application of devices appear to confirm that their use was widespread, at least common enough to draw censure. A wide variety of literature, from conduct books to dancing, health and medicine and gender, reinforced the importance of posture. Authors like Andry and Hogarth attempted to establish exact rules and proportions for the human body, perhaps by no coincidence at a time when classification and tabulation were at the heart of philosophical approaches to the natural world. But the true extent of consumption – and indeed regular use – is harder to discern.

What remains, however, is the clear willingness or *desire* to fashion the body. Once older ideas precluding people from interfering with their own body as the ultimate symbol of God's work had begun to decline, the material alteration of the body became acceptable, if not encouraged. As 'improvement' became entrenched as a concept across many areas of eighteenth-century life, the fabric of the body itself became a site of enhancement, correction and transformation.

Notes

1   Advertisement, 'Distortion', *True Briton*, 25 January 1800.
2   James Caulfield, *Blackguardiana: or a Dictionary of Rogues, Bawds, Pimps, Whores, Pickpockets, Shoplifters…* (London: 1793), 173.
3   Ibid.
4   Advertisement, *St James Chronicle or the British Evening Post*, 20 March 1764.
5   John Weaver, *Anatomical and Mechanical Lectures upon Dancing. Wherein Rules and Institutions for That Art Are Laid Down and Demonstrated* (London: printed for J. Brotherton and W. Medows, 1721), 94.
6   Georges Vigarello, 'The Upward Training of the Body from the Age of Chivalry to Courtly Civility' in Michel Feher (ed.), *Fragments for a History of the Human Body, Volume Two* (New York: Zone, 1990), 168–76.

DOI: 10.1057/9781137467485.0006

7   Lynn Sorge-English, *Stays and Body Image in London: The Staymaking Trade, 1680–1810* (London, 2011), 3–6.

8   David M. Turner, *Disability in Eighteenth-Century England: Imagining Physical Impairment* (London: Routledge, 2012); Helen Deutsch and Felicity Nussbaum (eds), *Defects: Engendering the Modern Body* (Ann Arbor: University of Michigan Press, 2000).

9   Turner, *Disability*, 22–5, 27.

10   Lennard J. Davis, 'Dr Johnson, Amelia and the Discourse of Disability in the Eighteenth Century' in Helen Deutsch and Felicity Nussbaum (eds), *'Defects': Engendering the Modern Body* (Ann Arbor: University of Michigan Press, 2000), 60.

11   William Hay, *Deformity: An Essay* (London: 1754), 4.

12   See Sorge-English, *Stays*, 138–9.

13   David M. Turner and Alun Withey, 'Technologies of the Body: Polite Consumption and Deformity in Eighteenth Century England', *History*, 99: 338 (2014), 781.

14   For example, Liliane Hilaire-Pérez and Christelle Rabier, 'Self Machinery? Steel Trusses and the Management of Ruptures in Eighteenth-Century Europe', *Technology and Culture*, 54:3 (2013), 460–502.

15   A term that, as Dorinda Outram argues, could bear a multiplicity of meanings: Dorinda Outram, *The Enlightenment* (Cambridge: CUP, 1995), 83.

16   Nicholas Andry, *Orthopædia: or, the Art of Correcting and Preventing Deformities in Children* (London: 1743), 45.

17   Timothy Sheldrake, *Essay*, 8.

18   Andry, *Orthopaedia*, 69, 71.

19   Ibid., 72.

20   Anon, *The Polite Academy, or School of Behaviour for Young Gentlemen and Ladies* (London: printed for R. Baldwin, 1762), 35–6.

21   Ibid., 36.

22   Anne Bloomfield and Ruth Watts, 'Pedagogue of the Dance: The Dancing Master as Educator in the Long Eighteenth Century', *History of Education*, 37:4 (2008), 607–8.

23   Ibid., 608.

24   Weaver, *Anatomical and Mechanical*, 81.

25   Andry, *Orthopaedia.*, 77.

26   William Hogarth, *The Analysis of Beauty: Written with a View of Fixing the Fluctuating Ideas Of Taste* (London: printed by J. Reeves, 1753), viii.

27   Ibid., 23.

28   Ibid.

29   Erasmus Darwin, *A Plan for the Conduct of Female Education, in Boarding Schools, Private Families, and Public Seminaries. By Erasmus Darwin, M.D. F.R.S* (Philadelphia: printed by John Ormrod, 1798), 113.

DOI: 10.1057/9781137467485.0006

30 Andry, *Orthopaedia*, 68.
31 Weaver, *Anatomical and Mechanical*, 83.
32 See, for example, Roy Porter, *Quacks: Fakers and Charlatans in Medicine* (Stroud: Tempus, 2003), especially ch. 2.
33 Sorge-English, *Stays,* 124; Turner and Withey, 'Technologies', 775.
34 Timothy Brand, *Chirurgical Essays on the Cure of Ruptures and the Pernicious Consequences of Referring Patients to Truss Makers* (London: 1785).
35 Advertisement, 'J. Whitford, Truss Maker to the City Truss Society for Relieving the Ruptured Poor, *The Morning Chronicle,* 4 August 1809; Advertisement, 'Double-Springed Elastic Truss', *The Oracle and Public Advertiser,* 10 March 1795; Advertisement, 'Ruptures Cured as Well as Palliated', *The Public Ledger,* 20 December 1765.
36 Brand, *Chirurgical Essays*, 41.
37 See for examples Advertisement, 'Guy Nutt still Liveth at the Sign of the White Naked Boy', *London Journal,* 23 February 1734; Advertisement, 'New Invented Patent Steel Spring Trusses', *Caledonian Mercury,* 1 April 1786.
38 Advertisement, 'The Way to Save Money and Prevent Robberies', *Oracle and Public Advertiser,* 6 March 1797.
39 Ibid.
40 Advertisement, 'Plain and Useful Instructions for the Relief and Cure of Ruptures, *The Critical Review or Annals of Literature,* 29 May 1800.
41 Liliane Hilaire Perez and Christelle Rabier, 'Self Machinery? Steel Trusses and the Management of Ruptures in Eighteenth-Century Europe', *Technology and Culture,* 54 (2013), 465.
42 British Library Patent number 3458, 15 June 1811, Specification for Obtaining Motive Power.
43 James Lucas, *A Candid Inquiry into the Education, Qualifications, and Offices of a Surgeon-Apothecary* (London: Stationers Hall, 1800), 92.
44 Advertisement, 'Dr De Malon Continues ...', *St James Chronicle or the British Evening Post,* 3 July 1773.
45 Advertisement, 'Ruptures Cured as Well as Palleated without Fear of Relapse', *Public Ledger,* 20 December 1765.
46 Turner, *Disability,* 131.
47 Ibid. For 'monstrosity', see Alan W. Bates, 'Good, Common, Regular and Orderly: Early Modern Classifications of Monstrous Births', *Social History of Medicine,* 18:2 (2005), 141–58; Lorraine Daston and Katherine Park, *Wonders and the Order of Nature, 1150–1750* (Boston: Zone Books, 1998); Dennis Todd, *Imagining Monsters: Miscreations of the Self in Eighteenth-Century England* (Chicago: University of Chicago Press, 1995).
48 Felix Wurtz, ' "An Experimental Treatise of Surgerie in Four parts... Whereunto is Added ... the Childrens Book", trans. Lenertzon Fox

DOI: 10.1057/9781137467485.0006

(1656)' in John Ruhrah (ed.), *Pediatrics of the Past* (New York: Paul B. Hoeber, 1925), 211. I am grateful to Hannah Newton for providing this reference.

49 James Cooke, *Mellificium chirurgiæ: or, the Marrow of Chirurgery* (London: printed for John Marshall, 1717), 236.

50 Andry, *Orthopaedia, Volume 1*, 82.

51 Sorge-English, *Stays and Body Image*, 113–16; Georges Vigarello, 'The Upward Training of the Body from the Age of Chivalry to Courtly Civility' in Michel Feher (ed.), *Fragments for a History of the Human Body, Volume Two* (New York: Zone, 1990, 168–76.

52 Advertisement, 'Meares, Truss maker, Ludgate Hill, *The Westminster Journal*, 4 February 1744.

53 Advertisement, 'Speedily Will Be Published', *Gazetteer and New Daily Advertiser*, 16 January 1768.

54 Advertisement, 'Rickets in Children Infallibly Cur'd', *Post Boy*, 3 January 1708; Advertisement, 'The Fam'd Worm-Plaister', *London Journal*, 30 May 1723.

55 John Allen, *Dr Allen's Synopsis Medicine* (London: printed for J. Pemberton, 1730), 221.

56 Andry, *Orthopaedia*, 127.

57 Timothy Sheldrake, *Observations of the Causes of Distortions of the Legs of Children* (London: Printed for Messrs Egerton, 1794), 1–8.

58 W.B., Letter to editor of *The Public Ledger*, London, 11 October 1765.

59 John Hill, *The Conduct of a Married Life: Laid Down in a Series of Letters Written by the Honourable Julia-Susannah Seymour, to a Young Lady*... (London: printed for R. Baldwin, 1754), 94.

60 Timothy Sheldrake, *An Essay on the Various Causes of the Distorted Spine* (London: printed for C. Dilly, 1783) 1, 2, 7.

61 Advertisement, 'To be had only of J. SLEATH', *The Oracle*, 28 April 1791.

62 Anon, *Village Memoirs: In a Series of Letters Between a Clergyman and his Family in the Country, and His Son in Town* (London: Printed for T. Davies, 1765), p. 137.

63 Anon, *The Pantheonites: A Dramatic Entertainment as Performed at the Theatre-Royal in Hay-Market* (London: Printed for J. Bell, 1773), 17.

64 Advertisement, 'The Way to Save Money and Prevent Robberies', *The Oracle and Public Advertiser*, 6 March 1797; Advertisement, 'Elastick Trusses for Ruptures, Bearing Down in Women &c', *The World*, 1 April 1790.

65 Advertisement, 'Change of Residence – Talmage's Warehouse', *The Morning Post*, 19 October 1811.

66 Anon, 'Concerning the Many Inducements for Coming to London', *Newcastle General Magazine*, 12 December 1753, 628.

67 Ibid.

68 There were debates, for example, about cosmetics and their role in concealment. See Morag Martin, *Selling Beauty: Cosmetics, Commerce and French Society, 1750–1830* (Baltimore: Johns Hopkins University Press, 2009),

DOI: 10.1057/9781137467485.0006

77; Jennifer M. Jones, *Sexing La Mode: Gender, Fashion and Commercial Culture in Old Regime France* (Oxford: Berg, 2004).

69 Advertisement, 'Steel Stays', *The Morning Herald and Daily Advertiser,* 18 April 1782.

70 Advertisement, 'Distorted Shapes in Ladies and Gentlemen', *London Evening Post,* 23 April 1778.

71 Ibid.

72 See the examples quoted in Turner, *Disability,* 121.

73 Advertisement, 'This is to Give Notice', *London Daily Post and General Advertiser,* 16 February 1739.

74 Sheldrake, *Observations,* 87.

75 For example see D. Merande, *A Succinct Account of a Machine, Newly Invented for the Cure of Praeternatural Curvatures of the Spine* (London: Printed for T. Jones, 1768), 7.

76 William Beckett, *Practical Surgery Illustrated and Improved* (London: Printed for F. Curl, 1740), 10.

77 Sheldrake, *Essay,* 22–3.

78 Jones, *Essay,* 45.

79 For studies on exercise in the eighteenth century, see Julia Allen, *Swimming with Dr Johnson and Mrs Thrale: Sport, Health and Exercise in Eighteenth-Century England* (Cambridge: Lutterworth, 2012); Rebekka von Mallinckrodt and Angela Schattner (eds.), *Sport and Physical Exercise in Early Modern Culture* (Farnham: Ashgate, 2016).

80 Deborah Lupton, *Medicine as Culture: Illness, Disease and the Body* (London: Sage, 2012 edition), 82–3

81 James McKittrick Adair, *An Essay on Regimen, for the Preservation of Health, Especially of the Indolent, Studious, Delicate and Invalid* (London: Printed by J&B Wilson, 1799), 67.

82 John Cheshire, *The Gouty Man's Companion; or a Dietetical and Medicinal Regime* (Nottingham: Printed by G. Ayscough, 1747), 44.

83 Francis Fuller, *Medicina Gymnastica; or Every Man His Own Physician* (London: printed for W. Norris, 1777), 31. See also James Tweedie, *Hints on Temperance and Exercise, Shewing their Advantage in the Cure of Dyspepsia, Rheumatism, Polysarcia and Certain Stages of Palsy* (London: printed by T. Rickaby, 1799).

84 Ibid.

85 Ibid., 266.

86 British Library, BL Patents 1779 no. 1211, 'Specification for Machines &C for Gymnastick Exercises'.

87 For example see Advertisement, 'The Great Secret of Curing Ruptures', *Courier,* 27 July 1793; see also Robert Thornton, *Medical Extracts. On the Nature of Health, with Practical Observations: And the Laws of the Nervous and Fibrous Systems* (London: Printed for Robinsons, 1795), 201.

DOI: 10.1057/9781137467485.0006

88   Jones, *Essay,* 46.

89   'Private Gentleman', *New Inventions and New Directions, Productive of Happiness to the Ruptured* (London: Sampson Low, 1800), 21.

90   For example, see Advertisement, 'By the King's Patent, Bath Elastic Saddles', *Bath Chronicle,* 30 April 1789. Maxfield was a truss maker.

91   Advertisement, 'Surgeons Instruments and Cutlery', *Adams' Weekly Courant, Chester,* 11 October 1744; Advertisement 'This Is to Give Notice', *London Daily Post and General Advertiser,* 16 February 1739.

92   Advertisement, 'Sir, Your Inserting This in Your Paper...', *General Advertiser,* 11 February 1749.

93   Advertisement, 'Surgeons Instruments'.

94   For example, Advertisement, 'Ruptures Cured as Well as Palliated', *The Public Ledger,* 20 December 1765.

95   Merande, *Succinct Account,* 7.

96   Ibid., 13.

97   Advertisement, 'The Great Secret of Curing Ruptures', *World,* 12 April 1793.

98   Advertisement, 'To be had only of J. Sleath', *The Oracle,* 28 April 1791.

99   Quoted in Joan Lane, 'The Diaries and Correspondence of Patients in Eighteenth Century England' in Roy Porter (ed.), *Patients and Practitioners: Lay Perceptions of Medicine in Pre-Industrial Society* (Cambridge: CUP, 1985), 232

100 Henry Manning, *Modern Improvements in the Practice of Surgery* (London: Printed for G. Robinson, 1780), 397.

101 Henry Heather Bigg, *The Gentle Treatment of Spinal Curvature* (London, 1875), 12.

102 Ibid., 13.

103 Merande, *Succinct Account,* 18.

DOI: 10.1057/9781137467485.0006

2
Shaving and Masculinity in Eighteenth-Century Britain

Abstract: *The eighteenth century was a period in which facial hair fell dramatically from favour. In this chapter, Withey provides a fascinating new study of the practice and importance of shaving in the eighteenth century. Often overlooked as a mundane task, shaving the face was invested with new meaning, as a smooth face came to symbolise openness, neatness and elegance. Underpinning this change was the introduction of new, sharper razors made of cast steel, sold increasingly sold by specialist makers who were innovators in metallurgy. Razor advertising increased markedly, appealing to a new market of men who shaved themselves, rather than visiting a barber. Advertisements appealed to masculine traits such as hardness, control and an interest in popular science. As Withey shows, the decline of facial hair was linked to various debates from gender and sexuality to medicine.*

Withey, Alun. *Technology, Self-Fashioning and Politeness in Eighteenth-Century Britain: Refined Bodies.*
Basingstoke: Palgrave Macmillan, 2016.
DOI: 10.1057/9781137467485.0007

As the previous chapter showed, control over bodily shape was a key element of politeness. Steel was an important component in that process. Equally important, however, was control over bodily surfaces, and perhaps especially the face. One practice that was certainly transformed by the introduction of cast steel was shaving. Razor making was a key area of production which benefitted from the new steel technology. As Chris Evans has recently suggested, the eighteenth-century cast steel razor and the newly domestic practice of shaving provide useful links between the concepts of 'Industrial Enlightenment' and the articulation of politeness.[1] Far from being on the periphery of any commercial revolution, as Neil McKendrick suggested, shaving and razors were in fact at its heart.[2]

This chapter proposes a new context for understanding shaving, one that foregrounds changes in masculinity and concepts of politeness and raises broader questions about individual experiences of self-fashioning. The practice of shaving draws upon earlier concepts of physical health and well-being, while the advertising strategies of razor makers dovetail with debates about the scope and boundaries of the medical marketplace. Also, the purchase of razors and shaving accoutrements offers insights into men's consumption, and masculinity. Something so apparently mundane as shaving actually embodies a broad spectrum of sociocultural and economic factors, as well as the role of steel in fashioning the body. Technological developments gave manufacturers a new material with which to enhance and embellish their goods. Razor makers also drove change, continually improving their products to try and outdo their competitors. The demands of enlightened self-fashioning called for better, more comfortable and even more aesthetically pleasing razors. Shaving was therefore enmeshed in a complex web of meanings, both born out of enlightened technological innovation and feeding back into the construction of polite bodily form. The practice of shaving was itself made easier by the introduction of a new technological innovation. But it would be simplistic to assume that new types of razors were the only impetus behind the new vogue for shaving during this period. Instead, new technologies interacted in complex ways with cultural influences.

Facial hair and masculinity in the eighteenth century

The relationship between eighteenth-century masculinity and politeness was itself complex. Indeed, concepts of both male behaviour and bodily

DOI: 10.1057/9781137467485.0007

appearance were in flux. Until fairly recently, however, men's bodies during this period were noticeably absent from discussions of sex and gender.[3] Some historians have argued that a sea change occurred in understandings of the gendered body during the Enlightenment. Thomas Laqueur's now widely debated arguments relating to supposed changes from a 'one-sex' to a 'two-sex' corporeal model raised new questions about eighteenth-century gender, the body, and sexual difference.[4] The later eighteenth century has been argued to be a period that witnessed the *formation* of new understandings of gender, which, for the first time, sited gender difference very firmly in the body.[5] Among the objections raised to Laqueur's work was a lack of a firm chronological framework for such changes, along with questions surrounding the social depth to which it penetrated. It could also be argued that, even in medical writings, matters like dress and education featured more than a clear model of masculine embodiment.

Matters are further complicated by a supposed change from politeness to sensibility in the last quarter of the eighteenth century. Philip Carter suggests that while the language and behaviours of sensibility and politeness were broadly similar, there was a shift away from the emphasis upon the outward display of politeness, and towards a more reflective and emotional demeanour.[6] Such changes coloured expectations of the male body. In the 1770s and '80s, a new and physically slight, even enervated, bodily ideal emerged. Men's bodies supposedly became more feminine and delicate as their feelings became more refined.[7] This was in contrast to previous emphases upon men's supposed physical superiority over women.

The gradual erosion of rigid terminologies of status from earlier periods saw terms such as 'gentleman' no longer strictly denoting patrician or gentry. Instead, 'gentleman' became elided with a social ideal of conduct – that is, behaviour and not status – and was therefore becoming more accessible to those of lower station.[8] Conduct literature sought to codify ideals of male behaviour and comportment, from language to demeanour. Men were encouraged to soften their countenance, refine their manners and engage women in stimulating conversation.[9] Even the polite dialogue of shop transactions conveyed character and status, and were particularly important where creditworthiness was at stake.[10]

In addition to his manners, a gentleman should 'look the part', by wearing the right clothes, in the right way.[11] Fashion items like wigs created a facade of sophistication and sartorial elegance, becoming almost

DOI: 10.1057/9781137467485.0007

ubiquitous. Wig wearing was generally viewed as suitably masculine, and grew in popularity lower down the social scale. This was another act that drew upon new shaving technologies, as heads were shaved in preparation for the wig.[12] But anxieties were raised about the emasculating effects of Frenchified fashions in Britain. As Penelope Corfield notes, the head represented authority, and the decision to cover or uncover it was socially loaded.[13] Wig wearing could thus be a sign of masculinity and an obvious symbol of dignity. But, as Michèle Cohen has suggested, wigs also represented a 'dilemma of masculinity'.[14] For some, elements of the new fashions appeared uncomfortably effeminate, especially when taken to the extreme in the figure of the macaroni. Wigs could convey learning, sagacity and means certainly. But they also blurred gender boundaries.

Shaving, however, seems to have avoided the taint of effeminacy that sometimes hung over the wig. Although shaving softened the countenance and, like a wig, effectively feminised a man's face, removing facial hair could imply control, discipline and self-mastery leading to, as Lord Chesterfield put it, 'harmonious self-presentation'.[15] Shaving opened up a man's face to the world, leaving nothing hidden from view. For John Clubbe's idealized young clergyman in 1765, 'openness of countenance is the characteristic of an ingenuous mind'.[16] This was a crucial point: 'openness' did not merely indicate a clear complexion but also symbolised a mind receptive to the new possibilities of the age. A freshly shorn face was the mark of the enlightened man, whereas to be facially hirsute spoke of savagery and wildness. By 1802, 'the caprices of fashion [had] deprived all the nations of Europe of their beards'.[17] After centuries of ubiquity, the beard was rendered undesirable within a mere hundred years. Some argued that the fashion for beardlessness was transient and undesirable. In 1789, a Frenchman, J.A. Dulaure, wrote to defend the beard, arguing that a thick crop of facial hair was the very mark of the enlightened European and a cause for admiration by members of other nations.[18] For Dulaure, citing images of ancient philosophers, the beard was a mark of wisdom that ennobled the face.[19] Shaving the chin was a 'disgraceful act', which made a man resemble a 'woman, a eunuch, or a child'. Worse still, 'a shaved chin was always a sign of slavery, infamy or debauchery'.[20] Sadly for Dulaure, most men apparently did not agree, and he was a lone advocate in a clean-shaven Europe.

Of central concern was that the body should be aesthetically pleasing to others. Something, though, was inherently *displeasing* about beards.

DOI: 10.1057/9781137467485.0007

In some respects this contradicts eighteenth-century cultural and physiological ideas about manhood. Male facial hair had been a basic marker of sexual difference. Along with genitalia, it was a defining physiological characteristic of a man, both in popular and scientific discourses.[21] It was also a key ethnic characteristic of Europeans that they were able to grow beards where other races could not. According to Charles White in 1799, only Europeans possessed noble characteristics such as the 'perpendicular face, prominent nose [and] that majestic beard'.[22] But it seems that the important element was the *ability* to grow a beard rather than necessarily wearing one. Karen Harvey cites the 1766 example of one Maria Brown, who judged the sexual differences between herself and a male acquaintance not only by his genitalia but also by his facial hair. Interestingly, however, although his beard was remarked upon, 'he shaved every day'.[23]

Why, then, did eighteenth-century men abandon facial hair? Identifying any specific turning point in early modern attitudes towards facial hair is problematic, but in the absence of firm evidence it is certainly worth some speculation. Moves towards a more delicate male appearance, noted above, are certainly one factor. But there are other possibilities, such as changing medical conceptions of the beard. In the early modern period, beards were central to medical conceptions of fertility and virility. As Will Fisher notes, facial hair in humoral medicine was directly linked to the production of semen, and was in fact a form of excreta directly resulting from heat arising from the testicles. In this sense the beard was a visible representation of the generative potential of a man.[24] In humoral medical theory, potentially harmful and excessive matter was driven out by evacuative measures. Shaving off stubble therefore rid the body of a potential source of sickness.[25] But, as the eighteenth century progressed, medical ideas were slowly changing. As mechanistic ideas of the body became more popular amongst the literati, beliefs in the humours were becoming associated with the 'vulgar'.[26] No evidence suggests a fixed point at which facial hair lost its humoral associations. This was instead a period of gradual transition as traditional concepts of the body and sexuality shifted. By 1800, though, it seems clear that a wet shave was done more for cosmetic than medical, reasons.

Cultural and ethnic difference may also have encouraged negative connotations of the beard. An interest in exotic races and cultures was certainly an important feature of polite society in the eighteenth century. Nevertheless, and contradicting White's letter, it is noticeable that removing the beard was taken as evidence of the superiority of

DOI: 10.1057/9781137467485.0007

Europeans over foreign cultures. Africans, for example, were bearded because they had no opportunity or inclination to shave, while culturally and technologically superior Europeans did.[27] The beard could reflect cultural 'otherness' in other ways. Angela Rosenthal highlights its use as a racial characteristic in satirical depictions of Jewish men.[28] It is no coincidence that Clubbe's *Letter of Free Advice to a Young Clergyman* also advised the subject 'not to come into that Jewish fashion of wearing a skirting of beard around the face'.[29]

A third possibility is the intrusion of aesthetics into bodily ideals. Greek writers had outlined rules of physiognomy and ideals of the face. These were revived in the eighteenth-century vogue for neoclassicism in art and architecture and, for some, such as Joshua Reynolds, acted as a virtual standard. It seems plausible to suggest that such standards were transposed onto ideals of physical appearance, although the fact that many Greek and Roman statues were bearded does not support this.[30] Nonetheless, there was certainly some precedent for the alteration of the body to meet ideals of a perfect, youthful body through the so-called 'cult of youth'. As romantic poets and enlightened thinkers lauded childhood as a state of perfect unspoilt nature, so some parents even began to deliberately undernourish their offspring, to cultivate a delicate appearance. Fear of ageing saw adults turn to cosmetic beauty washes and other artificial means to preserve the outward appearance of youth.[31] Shaving actually fits this framework quite neatly. The male child's face was smooth, and only in adolescence did it begin to display stubble. Removing facial hair thus rendered the face more youthful.

Lastly, it is possible that beards exemplified an opposing model of roughness and rugged masculinity. Facial hair for example was deliberately adopted by military regiments, and therefore had broader connections to imperialism and martial masculinity.[32] According to Christopher Duffy, certain French regiments insisted that recruits wear moustaches, even painting them on, or constructing false ones from horsehair, if the unfortunate recruit was unable to grow his own.[33] This outward demonstration of masculinity and virility sought to make enemies quail; it was no accident that tall men with moustaches were often placed at the head of a column.[34] As the male ideal became more refined, it is possible that such rough manliness fell out of favour.

Whatever the underlying causes, it is clear that beard wearing was seriously démodé by the mid-eighteenth century. Neither was being clean-shaven merely a British phenomenon; rather, it was Europe-wide.

DOI: 10.1057/9781137467485.0007

It is unclear whether the shaved face was a preoccupation of elites and middling sorts, and how far it filtered down the social scale. More work is needed on provincial sources, like barber accounts or poor law records, which is beyond the scope of this chapter. But certainly in elite circles, appearing unshaven or stubbly was undesirable, and in some cases even politically unwise, since it provided useful ammunition for political cartoonists.

One prominent figure who fell foul of satirists' pens was the Whig politician Charles James Fox. Fox was a swarthy man with a heavy beard – apparently even described as resembling a monkey by his father when a small child.[35] Cartoons nearly always depicted him with prominent stubble, but not a full beard. This was partly visual shorthand for his perceived earthiness and lack of refinement – he was often shown conversing with rough country types, as a 'man of the people'. His association with Georgiana Cavendish, the Duchess of Devonshire, however, added spice to satirical depictions. The Duchess, unusually given that she was not related to Fox, canvassed heavily for him, hosting tea parties and gatherings to garner support from society figures. A committed gamester, she was reputed to keep low company and repeatedly left her home and duties, leading to public accusations of impropriety and immorality.[36] In the public view, Georgiana was defeminized by her transgressions into a political world that was increasingly considered masculine, rather than feminine. Conversely, Fox (already seen as morally weak) was depicted as socially subservient, but also emasculated, his manhood diminished by the overpowering Duchess.

In some cases, the visual imagery was straightforward. One cartoon entitled *Carlo Khan's Triumphant Entry into St. Stephen's Chapel* depicted a stubbly Fox sitting atop the shoulders of the Duchess of Devonshire, proclaiming, 'I could never have got in without your Grace's assistance'.[37] Another entitled *Frith the Madman Hurling Treason at the King* shows a squab and swarthy Fox dressed in women's clothing, a comment on his reliance upon a woman.[38] A third, however, is even more intriguing and is reproduced in Figure 2.

In J. Moore's 1784 satire *The Political Shaver,* a subservient Fox shaves the Duchess of Devonshire. This image was carefully constructed to highlight the abandonment of conventional femininity by the Duchess, and her power over Fox. Although it is he who performs the traditionally male role of barber, gender roles are effectively reversed here to emphasise his weakness. More interesting perhaps is the fact that the *removal* of

DOI: 10.1057/9781137467485.0007

FIGURE 2   *Henry Kingsbury* The Political Shaver *(London: J. Moore, 1784)*
Source: Image copyright Lewis Walpole Library, used with permission.

facial hair is the significant element in this subversion. As the masculine figure, the Duchess has her beard shaved. Fox's stubble, by contrast, remains prominent, suggesting that the beard does not automatically confer manliness. Clearly this is one example, and its overt satire should not be writ large across eighteenth-century society. Nevertheless, the subversion of a distinctly male ritual does suggest the close link between shaving and masculinity, and the potential social implications of not removing facial hair.

Beards bore other connections to undesirable stereotypes or characteristics. Growing a beard could be a self-conscious act signifying withdrawal from society. John Wroe (1782–1863), leader of the 'Christian Israelites', suffering periodic symptoms of mania and developing radical religious ideas, shaved his head and grew a long, straggly beard to demonstrate his distrust of convention.[39] Artists and cartoonists seeking to display the dereliction of the body would likely endow a poor subject with a thick beard. Tim Bobbins' *Human Passions Delineated* (1773) used the beard, along with rickety teeth, to illustrate the rough earthiness of

DOI: 10.1057/9781137467485.0007

plebeian workers. James Gillray's *A New Way to Pay the National Debt* (1886), satirising the profligacy of the royal family, depicted a maimed veteran, notable by his lack of limbs and his unkempt beard.[40] Such depictions suggested loss of youth and virility. But the abandonment of personal grooming also acted more broadly as a metaphor for the rejection of fleshly appearance.

This is not to say that facial hair was always depicted or regarded pejoratively, nor universally outside Britain. Indeed, there was often ambiguity in portrayals of facial hair. Elderly subjects, especially in classical contexts, might be shown bearded. Thomas Jones's 1774 painting *The Bard* illustrates a poem by Thomas Gray. It depicts the suicide of a medieval Welsh bard, in the poem the last of his ilk, pursued to destruction by Edward I. The figure of the bard, in keeping with his surroundings of wild nature, was clothed in rags, his face adorned with a flowing beard. As easily as beards could connote the dereliction or decrepitude of age, they could also imply wisdom or gravity. Outside artistic representation, and in general, however, the removal of facial hair was socially, culturally and fashionably imperative.

The market for shaving products

As Amanda Vickery, Margot Finn and others have noted, male consumption in the eighteenth century is often overlooked.[41] Early consumption studies viewed women as the drivers of domestic acquisition, while men's desire for 'things' received less attention.[42] Finn has usefully demonstrated the increasing intrusion of men into the markets for both luxuries and household necessities.[43] Little attention has addressed, however, either the sorts of goods targeted specifically at men or the meanings conveyed by buying and owning them. Certain items were accoutrements expected of the modish gentleman, each with its own masculine connotations. Ownership of gold and silver watches increased massively after 1750, first in London then out into the provinces.[44] Watches were utilitarian items certainly, but also visible signifiers of prestige. As John Styles notes, however, they also spoke of a 'suitably masculine command of technology'.[45] The growing trend amongst elites for collecting scientific instruments could be seen as part of a similar discourse of male mastery over nature.[46]

Items of toilette for men were a growth area, signifying shifts towards a more individual male concern with personal grooming. By 1776, the

DOI: 10.1057/9781137467485.0007

London manufactory of Riccard and Littlefear offered many items especially for men, including 'gentlemens travelling cases' to enable business travellers and Grand Tourists alike to attend to their appearance while abroad. 'Roll-up pouches and etwee cases' provided space for 'all necessaries' from personal dress to writing equipment.[47] Others, like London perfumer Robert Sangwine in 1778, sold everything from shaving and dressing equipment designed to accompany gentlemen on sea voyages, to instruments allowing men to clean their own teeth.[48] Shaving paraphernalia fits this pattern well. Razor marketing affords a glimpse into the mechanisms of a uniquely male market, and the means used by makers and retailers to lure their customers.

It is firstly worth noting the important role of razor manufacturers as metallurgical innovators, and even acknowledged authorities on the composition of steel. The correspondence of Matthew Boulton, the eminent toy manufacturer and retailer, contains interesting evidence relating to the razor maker John Savigny of London. Letters to Boulton indicate that Savigny, 'a famous cutler in Pall Mall', was consulted about the composition of a sample of iron sent from overseas. Judging him as expert, Boulton was keen to learn Savigny's conclusions about the malleability, ductility and quality of the iron.[49] Patents for a new type of razor in 1789, by another member of the family, John Horatio Savigny, demonstrate the rigorous metallic standards often applied to these products. They were, he stated, only to be made 'of the purest steel which is to be forged (with very moderate and often repeated immersions in the fire) to ensure that the steel would receive no injury from a separation of its particles'. Savigny claimed that his unique method of tempering the steel gave his razors the 'most exquisite delicacy of edge'.[50] As with makers of postural devices, Savigny used his metallurgical knowledge to diversify, patenting inventions from a new type of tourniquet to stop bleeding, to a method for making and fixing ice skates with more ease, safety and expedition than hath hitherto been discovered'.[51]

Another prominent razor maker and cutler, James Stodart, was a Fellow of the Royal Society and published several essays relating to his comprehensive metallurgical experiments.[52] In 1821, he was an active, indeed sometimes senior, partner in a series of experiments on steel alloys conducted with Michael Faraday, but died before their completion. Faraday abandoned his own interest in alloying soon after his friend's death, leaving Stodart's contribution to the development of steel largely overlooked.[53] Newspaper advertisements for 'Peruvian steel' emphasised the importance of the work of 'Messrs Faraday and Stodart' whose experiments had led directly to the manufacture of the alloy.[54]

Cast steel offered new possibilities for razor manufacture, not least in the keenness of the edge. Razors made from older forms of steel existed throughout the eighteenth century, but the uneven carbon distribution rendered them liable to be brittle, easily blunted and uncomfortable to use. Constant stropping was necessary to maintain the cutting edge. Makers did their best to cry up the quality of these earlier products. In 1727, John Blanchard reassured the public that he had sourced, from 'Foreign parts ... a parcel of extraordinary good steel for razors', exceeding any expectations.[55] Others stressed the refinement of their products, instead of comfort or sharpness.[56] With cast steel razors, however, came a sharper edge and, according to newly converted manufacturers, less frequent sharpening. Amongst the first advocates of this new material was one Smith, a razor maker of Temple Bar in London, and maker of the 'Famous Polish'd Cast Steel Razors'. His self-aggrandising advertisements of 1765 claimed them to have been held in 'high esteem for many years', suggesting the new material was adopted almost as soon as it became available.[57]

A common tactic of razor makers was to appeal to elite and middling men's interest in scientific innovation and experimental philosophy. Many cried up their personal involvement in manufacture, and used sophisticated scientific tropes and the rhetoric of natural philosophy to embellish claims to the efficiency of their products. In 1788, Joseph Wright advertised his 'Philosophical Razors', which needed no setting or grinding, while another promised 'Concave Razors on Philosophical Principles'.[58] Stodart guaranteed the quality of his razors 'by prosecuting [scientific] experiments, as well as stamping his name on every blade.[59] An example of a typical straight razor made by Stodart can be seen in Figure 3. The use of cast steel was itself a selling point, and a slew of advertisements appeared in late eighteenth-century London newspapers for these new products. Form was often as important as function. Between October 1777 and December 1787, Robert Sangwine undertook a huge advertising campaign extolling the virtues of his 'cast steel razors' and 'polished razors', warranted to please.[60] John Palmer sold his 'high-polished cast-steel symmeter (scimitar) razors', stressing the highly reflective surface as well as the exotic connotations of the famed Turkish blade. No longer was the razor an item of utilitarian necessity; it could now be aesthetically desirable in its own right as part of the broader new vogue for steel goods.

DOI: 10.1057/9781137467485.0007

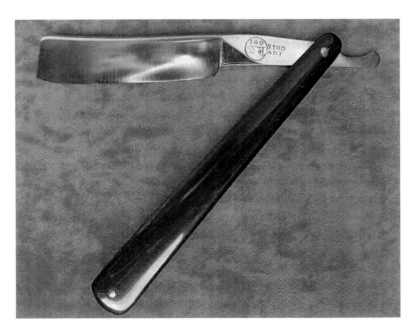

FIGURE 3   *Straight razor with manufacturers' mark of James Stodart*
Source: Image copyright Tony Holmes, http://www.Taylors1000.com, used with permission.

Razors were clearly (almost self-evidently) aimed at male consumers, but the rhetoric deployed by razor makers nonetheless appealed
to masculine ideals on a deeper level. In May 1791, Alexander Lowe of
Blackfriars, London, boasted that his method of working steel was of
'a certain degree of temper never known to fail'.[61] Palmer's razors were
rendered to a high degree of purity by hardening and tempering steel 'to
a certain degree of exactness'.[62] On a material level, emphasising durability was obviously important; the many razor strop advertisements
claiming to revitalise blunted blades attests to this. But the tempering
process was itself suggestive. Mastery over steel's physical toughness
symbolised masculine traits like hardness and control over the passions.
Other features of the razor suited masculine imagery. Its keen blade
bore obvious comparisons with the sword (sword making being closely
related to the crafts of cutlery and razor making) with all its attendant
martial connections. Men were encouraged to see razors as part of this
process of controlling their bodies.

DOI: 10.1057/9781137467485.0007

Ancillary equipment also contributed to the increasing marketization of shaving. From the 1750s, shaving was shifting from a prosaic daily task towards an overall experience. This was indeed a virtual new world of male pampering. Temple Bar hairdresser Richard Barnard, for example, claimed invention of the 'True Original Shaving Powder'.[63] The turner and toilet set maker James Emon sold shaving powder specially formulated 'to make razors cut easy and [was] very good for tender faces', alongside 'all necessaries for shaving'.[64] Perfumers such as Charles Lillie capitalised on a new vogue for post-shave care through products such as 'Persian (or Naples) Soap' to soothe smarting skin.[65] Even products such as 'Paris pearl water' and Mosenau's 'Essence of Cucumber', previously marketed to freshen women's skin and clear complexions, were now touted for men as ideal preservations for the skin after shaving.[66]

This nascent trade in shaving-related products even found form in furniture. From at least the late eighteenth century, elegant shaving stands appeared in furniture catalogues.[67] These items were functional, including storage space for equipment, a utility area and a mirror to monitor the act. But the appearance of the shaving table belied its prosaic function, being elaborately decorated with inlaid wood and ornate fittings. Here again, the emphasis was upon creating an *experience* of shaving, but also in some ways locating the physical act within wider constructions of the social expectations of male appearance.

The marketing of razors and shaving paraphernalia, therefore, proliferated during the later eighteenth century. As newer and better razors appeared, manufacturers deployed ever more sophisticated tropes to attract their customers. But the identity of these customers is a crucial point, and highlights fundamental shifts in male grooming routines during this period, at least in some levels of society. It is to questions of the experience and practice of shaving during the eighteenth century that the final part of this chapter turns.

The practice of shaving

Shaving the face was an intrinsically masculine act and one that, in the eighteenth century, changed from a practice almost exclusively done by professionals to one performed by a man upon himself, or by a servant. At the beginning of the century, blunt razors made it an uncomfortable experience. But, by its end, sharper cast steel blades transformed the process into part of the domestic toilette of a gentleman.

DOI: 10.1057/9781137467485.0007

Barbers of varying skill levels were widespread across eighteenth-century Britain and were the mainstay of shaving provision. In common with other trades, payment for goods and services was often on account, with the balance being settled biannually or annually. Receipts for Nottinghamshire barber Samuel Clowes offer an insight into the daily business of a provincial barber, including frequency. One Mr Evetts paid ten shillings for 'half years shaving' in April 1773, after Clowes charged him fifteen shillings for '1 years shaving' the year before.[68] In August the same year, Evetts paid seven shillings for shaving and wig dressing.[69] Evetts also patronised another Nottingham barber, William Sharp, whose accounts are more revealing of frequency. Between May 1768 and January 1769, Evetts visited Sharp on thirteen occasions to be shaved – roughly once every three weeks.[70] Whether every visit entailed shaving the face, head or both is unclear. But it seems unlikely that a man of status would wish to display an obviously stubbly appearance for most of each month. More plausible is that Evetts shaved himself far more regularly, perhaps visiting the barber at intervals for a more professional cut. The fact that the accounts reveal that Sharp sold a shaving brush to Evetts, strongly suggests that it was for Evetts' domestic use. Surviving 1766 accounts for the wealthy Warwickshire gentleman George Lucy of Charlecote suggest that he saw a barber at least weekly, again suggesting that he shaved at home in between.[71]

In other cases Sharp's accounts give some insight into prices, although it is difficult to generalise since, like physicians and apothecaries, barbers likely tailored their prices to their customers' means. Sharp's charge appears to have been between threepence and sixpence for a shave, with the former most frequent. There were also social and even legal conventions to be adhered to. It was technically illegal to be shaved on a Sunday, and occasional local prosecutions attest to the diligence of some communities in ensuring that this was enforced. In 1686, a Herefordshire barber, Edmund Hawley, was indicted not only for shaving customers in the time of service on a Sunday and Easter day but also for other offences including running a bawdy house and being a disturber of the peace.[72]

The figure of the inept barber was a comic staple, and satirical prints lampooned the misery caused to customers by the barber's clumsy efforts. The hapless subject in a 1799 Thomas Rowlandson print screams that he is about to lose his nose, as the country barber is absorbed by his own conversation.[73] The image in Figure 4 dates from 1804, and suggests the discomfort of the customer as the barber rasps his face with a blunt

DOI: 10.1057/9781137467485.0007

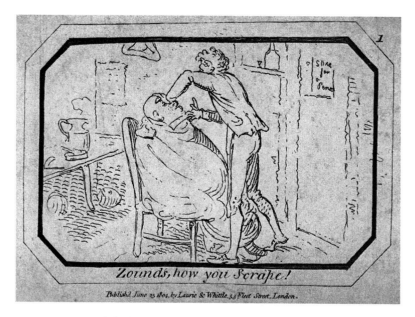

FIGURE 4   *Anon., 'A barber shaving a man in his shop' (London: Laurie & Whittle: 1804)*
Source: Image courtesy of Wellcome images.

razor. Common themes in such images were of the state of the barber's equipment, lack of skill and loss of concentration, leading to danger and potential injury.

The importance of being clean shaven carried beyond the domestic sphere and into institutions, including prisons and poorhouses, which commonly contracted barbers to shave their inmates. The Cumberland barber James Noble was employed by the county gaol to shave prisoners, but complained that after two years that he had yet to be paid.[74] Similar arrangements existed elsewhere. In the Bedfordshire county gaol in 1801, prisoners were shaved every week at a cost of one shilling and sixpence.[75] In Meirionydd in 1814, authorities paid one Daniel Jones for tasks including washing and shaving the prisoners.[76] Clearly, in such cases, politeness was not a consideration, but control and presentation were still apposite. Authorities imposed a disciplinary routine of cleanliness to emphasise their control over prisoner's bodies.

The barber (himself a member of a profession in transition as barbers split from surgeons to form their own company in 1745) doubtless

DOI: 10.1057/9781137467485.0007

remained an important figure in the provision of shaving and other bodywork.[77] It is not intended here to suggest that barbers were entirely circumvented during the eighteenth century, nor indeed that they did not still provide the main source of shaving. Razor advertising, however, suggests that responsibility for cosmetic appearance was shifting from the barber's shop to the home.

In October 1768, an advertisement was placed in the *Gazetteer and New Daily Advertiser* by William Riccard, a London cutler and ware-houseman. It was primarily an unremarkable puff for Riccard's newly invented instrument for animal venesection, but also mentioned another of his products – cast steel razors. Riccard was not the first to manufacture such articles; they were widely available from many outlets. But, the wording of his advertisement was new. His razors were targeted at a very specific stratum of society – 'for the ease and convenience of *those gentlemen who shave themselves*' [my emphasis].[78] In 1774, Daniel Cudworth was similarly advertising his newly invented razor strop 'to all gentlemen who shave themselves'.[79] J. Peney of London, a perukemaker, was less discerning, targeting his new razor strap to '[a]ll Gentlemen *or others* [my emphasis] who shave themselves'.[80] Who these 'others' might have been is debatable. One interpretation is the increasing intrusion of affluent middling sorts into this new market. By 1809, there were other signs that the barber no longer monopolised shaving. The anonymous article writer 'P' described a village barber, Joe Barrington, an exemplar of a solid British character, who knew little about French fashions, wigs or scents, but regaled his customers with earthy anecdotes as he shaved them. Even the misty-eyed 'P' had to concede however that Barrington was an 'unpolished remnant' of a bygone village barber's age and, in the last analysis, only 'old fashionedly useful'.[81]

How far the trend for autopogonotomy (self-shaving) was repeated across Georgian society is difficult to gauge. Individual shaving routines are obscure, especially since diaries and letters tend to note shaving only where mishaps occur. The diarist James Woodforde noted cutting his fingers while shaving too hurriedly in 1798.[82] It is likely that shaving was a regular activity for the majority of middling and elite men by the late eighteenth century, although how regular is unclear. If a razor was available at home, it seems plausible that a polite gentleman would tolerate no more than a few days' stubble at most. For those of means, this trans-lated as hiring servants capable of doing the job for them. For young men, such as 'T.H.', desiring a place as a gentleman's butler or valet, the

DOI: 10.1057/9781137467485.0007

ability to 'shave and dress' a potential master was worth stressing, and he was typical of many who did so, taking out an advertisement in the *Daily Advertiser* in 1773.[83] But, without the requisite skills in honing and wielding a razor, shaving oneself could undoubtedly be uncomfortable – something that strop manufacturers sought to capitalise upon.

The shaved face appeared to be a standard to which all men should aspire. In the later eighteenth century, popular accounts attested to the often-astonishing abilities of severely impaired or disabled people. Shaving was often among the tasks attributed to these people. One Thomas Pinington reputedly shaved himself despite having no hands, feet or legs, as did John Sear of London. William Kingston of Somerset had no arms, but used his feet for everything from shaving to boxing.[84] Handling a lethally sharp blade without injuring oneself was hard enough, but to do so with severe impairments required incredible dexterity. Doubtless some accounts were carefully constructed to titillate. But they also highlight shaving as a basic marker both of hygiene and capacity. By demonstrating their ability to perform it unaided, such men indicated that not only could they attend to their own personal toilette but also perform the tasks expected of a man.

Self-shaving was a relatively new practice, and one in which many men were unpractised and unprepared. If the clumsy barber was culpable, many individuals also doubtless injured themselves with their own clumsy efforts. A growing literature sought to educate men in shaving themselves quickly, cleanly and expeditiously. Again drawing upon the perceived superiority of the knowledge of razor makers, the first such book was that of a Parisian cutler, Jean-Jacques Perret (1730–1784). Perret's *Pogonotomie, ou l'Art d'Apprendre à se Raser Soi-Même* (Pogonotomy, or the art of shaving oneself) was first published in 1769, and instructed men how to angle a razor for best cutting and to avoid irritation, whilst also effectively inventing the safety razor. In 1776, J.H. Savigny published his own *Treatise on the Use and Management of a Razor*, which unsurprisingly stressed the metallurgical technologies of his razors and how to maintain them, as well as the process of shaving itself.[85] By the mid-nineteenth century, there were many dedicated volumes advising on everything from the condition and maintenance of razors to the need to use hot, instead of cold, water when shaving. Given that the mid-Victorian period witnessed a startling rise in the popularity of beards, it is likely that demand for books about autopogonotomy (self-shaving) fell just as swiftly.[86]

DOI: 10.1057/9781137467485.0007

Conclusion

This chapter has placed shaving at the heart of social, cultural and technological change in the eighteenth century. As other chapters in the book seek to demonstrate, the creation of self, and especially the polite self, in the eighteenth century relied not only upon manners and conduct but also upon appearance, shaped through the ownership and display of consumer goods. Where polite accoutrements have entered studies of politeness, they have most usually been exogenous – that is, things *on* the body, rather than alterations to the corporeal fabric itself. Shaving enabled men to alter their appearance to conform to a social ideal, and the introduction after 1750 of sharper, more durable steel razors made closer and more comfortable shaves possible. This is not to say, however, that technology was necessarily the driving force behind the eighteenth-century flight from the beard. As this chapter has suggested, shaving incorporated a number of interweaving and overlapping discourses about appearance, status, gender, respectability and health, as well as masculinity.

As debates about wigs demonstrate, the acceptable boundaries of manliness were mutable. Wig wearers expected to display learning, sagacity and male authority, but wigs might also suggest effeminacy, when in extreme forms. Despite the fact that shaving rendered the male face smoother and more feminine, such connotations seemingly bypassed shaving and *being* clean shaven. Unlike wig wearing, too, shaving the face was almost self-evidently an exclusively male activity. Indeed, there was an ever-increasing range of toilette products for men, highlighting popular acceptance for this form of male grooming.

Secondly, it seems apparent that the reaction against beards was widespread, although caution must be exercised here, for the sources (newspapers, satirical prints, portraits) all favour the upper and middle classes. It is entirely possible that the lower orders wore beards, perhaps through necessity but equally through choice. It is unsafe to assume that 'ordinary' people simply aped their social superiors. Nonetheless, the apparent fall of beards is suggestive of a society in which male facial hair had become socially undesirable.

Thirdly, as the discussion on the market for shaving products has shown, advertisers drew on strongly masculine discourses within their advertising. Shaving technology lent itself well to male hardness and strength – a fact still exploited by razor manufacturers today. Appeals to acceptable male interests such as science linked shaving to the latest technological

DOI: 10.1057/9781137467485.0007

innovations, as well as the cachet of owning an item made from desirable new materials, one that was aesthetically pleasing as well as utilitarian.

The treatment of facial hair and shaving as metaphors in cartoons and images suggests the existence of popular stereotypes. The use of shaving in the political satires relating to Charles James Fox, for example, while by no means definitive, suggest the strong link between the shaving and masculinity. Likewise comic depictions of barbers, and their hapless customers, highlighted what was probably an uncomfortable, indeed painful, activity for most men. There was a distinct change over time, however, as men slowly began to eschew the time and expense of the barber in favour of doing the job themselves. The growing number of self-help books, together with the burgeoning market for shaving accoutrements and types of razors, further attest to this change in habits – a change which was to last more than a century.

Finally, the market for shaving equipment was a uniquely male area of consumption. Shaving products exemplified changing male attitudes towards personal grooming during this period. Men were encouraged to pay more intimate attention to their own bodies, as well as their appearance. Something so apparently mundane as shaving can therefore be viewed as central to understanding not just the practice and purpose of male toilette but also the wider process of men's participation in a burgeoning Georgian consumer market.

Notes

This chapter is adapted from 'Shaving and Masculinity in Enlightenment Britain', *Journal of Eighteenth-Century Studies*, 36:2 (2013), 225–47, and is used by kind permission of the editor.

1   Chris Evans, 'Crucible Steel as an Enlightened Material', *Historical Metallurgy*, 42:2 (2008), 81.

2   Neil McKendrick, 'George Packwood and the Commercialization of Shaving: The Art of Eighteenth-Century Advertising or "the Way to Get Money and Be Happy"' in Neil McKendrick, John Brewer and J.H. Plumb (eds), *The Birth of a Consumer Society: The Commercialization of Eighteenth-Century England* (London: Europa Publications, 1982), 147.

3   Karen Harvey, '"The Majesty of the Masculine Form": Multiplicity and Male Bodies in Eighteenth-Century Erotica' in Tim Hitchcock and Michèle Cohen (eds), *English Masculinities, 1660–1800* (London: Longman, 1999), 193–6.

DOI: 10.1057/9781137467485.0007

4   Thomas Laqueur, *Making Sex: Body and Gender from the Greeks to Freud* (Harvard: Harvard University Press, 1990). For critiques of Laqueur's work, and others, see Dror Wahrman, 'Change and the Corporeal in Seventeenth and Eighteenth-Century Gender History: Or, Can Cultural History Be Vigorous?', *Gender and History*, 20:3 (2008), esp. 584–90.

5   Laqueur is an obvious example, but also see Outram, *The Enlightenment* (Cambridge: CUP, 1995), especially ch. 6; Anne Vila, 'Medicine and the Body in the French Enlightenment' in Daniel Brewer (ed.), *The Cambridge Companion to the French Enlightenment* (Cambridge: CUP, 2014), 199–213.

6   Philip Carter, *Men and the Emergence of Polite Society, Britain, 1660–1800* (Harlow: Pearson, 2001), 28–9. For different perspectives on sensibility, including its criticisms and decline, see G.J. Barker-Benfield, *The Culture of Sensibility: Sex and Society in Eighteenth-Century Britain* (Chicago: University of Chicago Press, 1996); Paul Goring, *The Rhetoric of Sensibility in Eighteenth-Century Culture* (Cambridge: CUP, 2004).

7   Carter, *Emergence*, 100, 105.

8   Klein, 'Politeness', 876.

9   Michèle Cohen, *Fashioning Masculinity: National Identity and Language in the Eighteenth Century* (London and New York: Routledge, 1996), 9.

10   Helen Berry, 'Polite Consumption: Shopping in Eighteenth-Century England', *Transactions of the Royal Historical Society*, 12 (2002), 388.

11   Klein, 'Politeness'. 882.

12   Michael Kwass, 'Big Hair: A Wig History of Consumption in Eighteenth-Century France', *American Historical Review*, 111:3 (2006), 634.

13   Penelope Corfield, 'Dress for Deference and Dissent: Hats and the Decline of Hat Honour', *Costume*, 23 (1989), 71.

14   Cohen, *Fashioning Masculinity*, 9.

15   Quoted in ibid., 45.

16   John Clubbe, *A Letter of Free Advice to a Young Clergyman* (London: 1765), quoted in John Adey Repton, *Some Account of the Beard and the Moustachio, Chiefly from the Sixteenth to the Eighteenth Century* (London: J.B. Nichols, 1839), 6.

17   William Nicholson, 'Philosophical Discquitions (sic) on the Processes of Common Life: – Art of Shaving', *A Journal of Natural Philosophy, Chemistry and the Arts,* Volume 1 (1802), 47.

18   J.A. Dulaure, *Pogonologia: or a Philosophical and Historical Essay on Beards, translated from the French* (Exeter: Printed by R. Thorn, 1789), 13.

19   Dulaure, *Pogonologia*, 26.

20   Ibid., 34.

21   Karen Harvey, *Reading Sex in the Eighteenth Century: Bodies and Gender in English Erotic Culture* (Cambridge: Cambridge University Press, 2004), 95–6.

22   Quoted in Angela Rosenthal, 'Raising Hair', *Eighteenth-Century Studies*, 38:1 (2004), 2.

DOI: 10.1057/9781137467485.0007

23 Harvey, *Reading Sex,* 96.

24 Will Fisher, 'The Renaissance Beard: Masculinity in Early Modern England', *Renaissance Quarterly,* 54:2 (2001), 173–5.

25 Sandra Cavallo, *Artisans of the Body in Early Modern Italy: Identities, Families and Masculinities* (Manchester: Manchester University Press, 2007), 39–40.

26 Porter, *Flesh in the Age of Reason* (London: Penguin, 2003), 60, 246.

27 Rosenthal, *Raising Hair,* 3.

28 Ibid., p. 3.

29 Clubbe, *Letter of Free Advice,* 6.

30 Porter, *Flesh in the Age of Reason,* 246.

31 Ibid., 242–3.

32 Christopher Oldstone-Moore, 'The Beard Movement in Victorian Britain', *Victorian Studies* 48:1 (2005), 11–13.

33 Christopher Duffy, *The Military Experience in the Age of Reason* (London: Routledge & Kegan Paul, 1987), 103.

34 Ibid., 111; Carolyn D. Williams, 'Half a Charge and No Wadding': Women and Guns in the Eighteenth Century', *Journal for Eighteenth-Century Studies,* 25:2 (September 2002), 249.

35 ODNB online – Charles James Fox

36 Phyllis Deutsch, 'Moral Trespass in Georgian London: Gaming, Gender and Electoral Politics in the Age of George II', *Historical Journal* 39:3 (1996), 648, 651–2.

37 Samuel Collings, '*Carlo Khan's Triumphant Entry into St. Stephen's Chapel*' (London: 1784).

38 I. Cruickshank, *Frith the Madman Hurling Treason at the King* (London: 1790).

39 See entry on John Wroe in Oxford Dictionary of National Biography, available online at http://www.oxforddnb.com/public/articles/13135851724383079.html, accessed on 2 September 2014.

40 See images and discussions in Diana Donald, *The Age of Caricature: Satirical Prints in the Reign of George III* (New Haven and London: Yale University Press, 1996), 10, 68.

41 Amanda Vickery, 'His and Hers: Gender, Consumption and Household Accounting in Eighteenth-Century England', *Past and Present,* Supplement 1 (2006), 12–38; Margot Finn, 'Men's Things: Masculine Possession in the Consumer Revolution', *Social History,* 25:2 (2000), 133–55; Maxine Berg, *Luxury and Pleasure in Eighteenth-Century Britain* (Oxford: Oxford University Press, 2005), 242.

42 Finn, 'Men's Things', 134.

43 Ibid., 138–42.

44 Styles, *The Dress of the People,* 97.

45 Ibid., 107.

DOI: 10.1057/9781137467485.0007

46   Amanda Vickery, *Behind Closed Doors: At Home in Georgian England* (New Haven and London: Yale University Press, 2009), 264–5

47   Advertisement for Riccard and Littlefear's manufactory, *London Evening Post*, 9–11 July 1776

48   Advertisement for 'Cast Steel Razors' by R. Sangwine, *St James Chronicle or the British Evening Post*, 7–10 February 1778.

49   Birmingham Archives, Matthew Boulton papers, MS 3782/12/23/33, Letter from William Allen (London) to MB, 2 April 1764.

50   British Library Patent Records, Patent Number 1716, Specification of John Horatio Savigny, 8 December 1789.

51   British Library, Patent Records Patent Number 2387, Specification of John Horatio Savigny, 31 March 1800 and Patent Number 1458, Specification of John Henry Savigny, 4 December 1784.

52   For examples, see James Stodart, 'On an Experiment to Imitate the Damascus Sword Blade', *Journal of Natural Philosophy, Chemistry and the Arts*, VII (1804), 231–2; J. Stodart and Michael Faraday, 'Experiments on the Alloys of Steel, Made with a View to Its Improvement', *Quarterly Journal of Science, Literature and the Arts*, 9 (1820), 319–30; Michael Faraday, 'An Analysis of Wootz, or Indian Steel. By M. Faraday, Chemical Assistant to the Royal Institution, *Quarterly Journal of Science*, 7 (1819), 288–90.

53   Sir Robert Hadfield, *Faraday and his Metallurgical Researches, With Special Reference to Their Bearing on the Development of Alloy Steels...*(London: Chapman and Hall, 1931), 39–42.

54   See *Leeds Mercury*, Saturday 30 September 1826.

55   Advertisement, 'John Blanchard, Razor Maker', *Mist's Weekly Journal*, 8 April 1727.

56   For example, Advertisement, 'All Gentlemen are Desir'd to Take Notice', *Daily Advertiser*, 22 October 1743.

57   Advertisement, 'Smith, Razor Maker, Six Doors from Temple Bar', *Gazetteer and New Daily Advertiser*, 21 February 1765.

58   Advertisement, J. Palmer, 'At no. 40 Castle Street ...', *World*, 23 February 1790; Anon, *St. James Chronicle or the British Evening Post*, 13 January 1781.

59   See Advertisement, 'R. Hawkesley at the Lilley ..', *World and Fashionable Advertiser*, Tuesday, 23 October 1787; Advertisement, 'J. Palmer, No. 40 Castle Street ...' *World*, Tuesday, 23 February 1790; Advertisement, 'Lowe's Cast Steel Razors', *Oracle*, Friday, 6 May 1791; Advertisement, 'Joseph Wright', *Bath Chronicle*, 10 January 1788; Advertisement, 'An Improvement in Cutlery', *World*, Saturday, 16 August 1788.

60   See, for example, Advertisement, R. Sangwine, 'Cast Steel Razors ...', *Morning Chronicle and London Advertiser*, 28 October 1777.

61   Advertisement for Alexander Lowe, 'Razors Warranted', *The Oracle*, 6 May 1791.

DOI: 10.1057/9781137467485.0007

62 Advertisement for 'J. Palmer, no. 40 Castle Street ...', *World,* 23 February 1790.
63 Advertisement, 'To Be Sold by Richard Barnard ...', *Daily Advertiser,* 16 January 1752.
64 Advertisement 'J. Emon's New Invented Powder ...', *London Evening Post,* 22 February 1752
65 Advertisement 'Persian (or Naples) Soap ...', *Daily Advertiser,* 22 November, 1744.
66 Advertisement 'Imported from Paris', *Public Advertiser,* 27 January 1756; Advertisement, 'To the Ladies', *London Courant and Westminster Chronicle,* 13 March 1780.
67 Vickery, *Behind Closed Doors,* 284–5.
68 Nottinghamshire Archives, MS DDE 98/187, Receipt for monies due from 'Mr Evetts' to Samuel Clowes, 15 April 1773; Nottinghamshire Archives, MS DDE 98/181, Receipt for monies due from 'Mr Evetts', Month unclear, 1772.
69 Nottinghamshire Archives, MS DDE 98/192, Receipt for monies due from 'Mr Evetts to Samuel Clowes, 26 August 1773.
70 Nottinghamshire Archives, MS DDE 98/105, Receipt for monies due from 'Mr Evetts' to William Sharp, 23 January 1769.
71 Warwickshire Record Office, MS L6/1320, Receipt bill from William Orchard to George Lucy esq., 11 May, 1766.
72 Herefordshire Record Office, MS BG11/17/5/72, Details of Charges against Edmund Hawley, 1686.
73 Thomas Rowlandson, *Barber Woodward* (London: Printed by T. Akerman, 1799).
74 Cumbria Record Office, MS WQ/SR/403/14, Petition of James Noble to Quarter Sessions at Appleby, Michaelmas Sessions, 1777.
75 Bedfordshire and Luton Archives and Record Service, MSS QSR/18/1801/126,142,151,154, Gaoler's general bills for county gaol, various dates, c. 1801.
76 Gwynedd Archives, Meironydd Record Office, MSS ZQS/E1814/6, Meirionydd county treasurer's accounts, 17 April 1814.
77 See, for example, Margaret Pelling, *The Common Lot: Sickness, Medical Occupations and the Urban Poor in Early Modern England* (London: Longman, 1998); Margaret Pelling, 'Medical Practice in Early Modern England: Trade or Profession?' in Wilfrid Robertson Prest (ed.), *The Professions in Early Modern England* (1987), 90–128.
78 Advertisement, 'A New Invented Instrument for Bleeding Horses', *Gazetteer and New Daily Advertiser,* 6 October 1768.
79 Advertisement, Daniel Cudworth, 'To All Gentlemen Who Shave Themselves', *Public Advertiser,* 14 April 1774.
80 Advertisement, 'To All Gentlemen or Others Who Shave Themselves', *General Advertiser,* 18 April, 1752.

DOI: 10.1057/9781137467485.0007

81   'P', 'Joe Barrington: Tub Village Barber' in *The Olio: Museum of Entertainment*, Volume 2, (1804).

82   John Beresford (ed.), *The Diary of a Country Parson: The Reverend James Woodforde, Vol. V, 1797–1802* (Oxford: Clarendon Press, 1968 ed.), 140.

83   Advertisement (untitled), *Daily Advertiser* (London), 26 April 1773.

84   Daniel Lysons, *The Environs of London*, Volume 1 (London: T. Cadell and W. Davies, 1891), 533.

85   J.H. Savigny, *Treatise on the Use and Management of a Razor: with Practical Directions Relative to Its Appendages* (London: Printed by T. Bensley, 1776).

86   For examples, see E. Rhodes, *Essay on the Manufacture, Choice and Management of a Razor by E. Rhodes, Cutler of Sheffield* (Sheffield: G. Ridge, 1824); Anon, *The Gentleman's Companion to the Toilet or a Treatise on Shaving, by a London Hair Dresser* (London: W. Strange, 1844).

DOI: 10.1057/9781137467485.0007

3
Managing the Body: The Material Culture of Personal Grooming

Abstract: *In an innovative study of a number of small instruments, Withey offers a new perspective on the importance of personal grooming in eighteenth-century Britain. As new attention focussed on the appearance of bodily surfaces such as faces, hands and teeth, so grooming instruments achieved new significance. These ranged from luxurious 'toilet sets' and 'equipages', and expensive, fashionable cases, to the smallest instruments such as nail nippers, scissors and tweezers. Steel was an important component in the manufacture of these goods, but, as Withey demonstrates, the act of attending to appearance was an important part of the conveyance of politeness and civility in the public sphere. Indeed, in many cases, instruments were routinely carried about the person, to allow people to refine their appearance on the move. Drawing on sources from contemporary literature to court records, Withey demonstrates how even the most basic and prosaic of artefacts can be loaded with historical significance.*

Withey, Alun. *Technology, Self-Fashioning and Politeness in Eighteenth-Century Britain: Refined Bodies.* Basingstoke: Palgrave Macmillan, 2016. DOI: 10.1057/9781137467485.0008.

> *Good health and longevity depends much upon personal cleanliness, and a variety of habits and customs, or minute attentions that it is impossible here to discuss. It were much to be wished that some author would undertake the trouble of collecting the result of general experience upon the subject, and would point out to those habits which, when taken singly, appear very trifling, yet when combined there is every reason to believe that much additional health and comfort would arise from their observance.*[1]

Sir John Sinclair's comment from his 1802 *Essays on Health and Longevity* refers indirectly to the importance of personal grooming in the eighteenth and early nineteenth centuries. The 'habits...customs, or minute attentions' to which he refers doubtless encompass the many routine tasks of daily maintenance that people were inclined to perform. This might include everything from combing hair for nits to the prompt emptying of chamber pots, but equally to the prosaic daily routines of appearance. On their own, things like plucking eyebrows or cutting nails 'appear very trifling'. Only in aggregate, Sinclair suggested, could their full importance be appreciated. But he also obliquely raises another problem – how to access these habits. Since individuals seldom recorded their own grooming routines, the task set for his imaginary author of recording the 'general experience' is a difficult one. While many conduct and etiquette books referred to the importance of keeping bodies clean and tidy, they seldom expanded on how to do it. People rarely troubled to write about how well they had cut their nails or plucked their eyebrows. Newspapers were not full of advertisements from tweezers makers touting their new, technologically advanced products. And yet personal grooming achieved new importance in the eighteenth century, and small instruments were central to this process. Whilst other chapters have explored individual instruments in detail, this chapter takes a different approach in looking at a number of different instruments in the context of body areas and surfaces. Unlike other discussions, in which evidence for usage, advertising and manufacture is more abundant, the focus here is upon what might be termed the 'ghost' of demand, consumption and use found in evidence for broad changes in attitude towards bodily maintenance, and certain bodily features, as well as the design, manufacture and retailing of objects themselves.

DOI: 10.1057/9781137467485.0008

The eighteenth century brought great changes to patterns of consumption. A new commercial society, changes in manufacturing and working methods all contributed to a 'material transformation of plebeian life'.[2] Whether or not this represents the 'commercial revolution' championed by Neil McKendrick is debatable, but consumer goods, including necessities and luxuries, were bought in increasing numbers across the social spectrum. By the 1770s, foreign observers were remarking upon the English obsession with buying and, by the 1790s, the extent of their spending was raising concern.[3] Much attention has naturally focussed upon the goods filling Georgian houses. Everything from tea sets and ornaments to furniture and wallpaper, all reflected the latest tastes for the exotic and the oriental.[4] From large country houses to the parlours of affluent middling sorts, and even in more modest plebeian dwellings, self-presentation, and even self-image, was bound up with the ownership and display of luxurious goods. The new penchant for spending extended to self-fashioning, and historians have explored a number of ways in which this supposed spending boom impacted upon the body.

One area that has received surprisingly little attention, however, is personal grooming. Cosmetics and bodily cleanliness and hygiene have certainly been explored in the context of bodily management and beautification. A huge range of cosmetics was available, in part linked to a 'cult of femininity' in which a smooth, clear complexion was valued, although ambivalence remained about the dangers of vanity.[5] Morag Martin has charted both the popularity of cosmetics in eighteenth-century France, and the religious, intellectual and sexual debates surrounding beautification.[6] Martin argues that, in a process of medicalization of the female body, doctors claimed dominion over cosmetic use, arguing that it was deleterious to health.[7] As Virginia Smith has noted, the late eighteenth century in Britain brought changes in attitudes towards personal hygiene. While medical authors had long advocated the removal of excrementitious matter and bodily pollutions, the period especially after 1770 saw these ideas become transmuted or incorporated into new concepts about bodily cleanliness – what Smith terms a 'civil cleanliness'.[8] But the means through which this was achieved, the actual instruments of personal grooming, have received little attention. This chapter contends that such small, quotidian items were actually important vectors through which people met social expectations of neat and elegant appearance.

The second half of the eighteenth century saw an increase in the availability and use of small instruments for the body. Unlike goods such

DOI: 10.1057/9781137467485.0008

as deportment devices, discussed earlier, it was not that new types of instrument were necessarily invented, although technology did bring material improvements to some. Rather, quotidian goods such as nail clippers, tweezers and toothpicks were imbued with new significance. Such items have generally been subsumed within the broader category of 'toys' – small, cheap goods of various sorts available in increasing numbers, often from wholesale retailers. As such they are regarded as fashion goods similar to buttons, seals and trinkets, rather than items of bodywork, or indeed medicine. It is perhaps easy to assume that they have no place in the history of health or the body. To do so, however, downplays their importance in the purposeful construction and management of the body and self. Items such as tweezers and toothpicks were part of a continuum of metallurgical knowledge that increased both the availability and utility of goods, and also the manual dexterity and skill of consumers.[9]

True, the use of small instruments was neither new nor unique to the eighteenth century. But there were changes, not least those brought about by the devolution of responsibility for bodily care from medical practitioners and towards the individual.[10] As seen in the previous chapter, Barbers (as barber-surgeons) traditionally bore responsibility for quasi-medical tasks like shaving, and reinforced the role of the practitioner in bodily maintenance, as part of broader health regimens. After the barbers' company split from the surgeons in 1745, barbering slowly became less 'medical' and more focussed on hairdressing and shaving. Whilst barbers were still important in the management of the body, individuals began to take more of a personal stake in their own appearance.

As David Turner has noted, the eighteenth century witnessed a cultural and commercial obsession with the 'body beautiful.'[11] As the body came to be understood in terms of refinement and harmonious appearance, small instruments became prominent in the daily management of appearance and self. Their ownership and use were further linked to the commoditisation and commercialisation of bodily care. Bodily routines were yoked to the desire to own objects, which, in turn, were imbued with the owner's expectations of self-fashioning. Such items both satisfied the vanity of consumers and also reflected a desire to display modernism and innovation.[12] In this sense, instruments were progenitors of a neat, 'polite' body. This process was certainly considered by contemporaries such as the economist Adam Smith, who referred to

DOI: 10.1057/9781137467485.0008

the increasing consumption of 'trinkets of frivolous utility' including toothpicks, ear scrapers and nail scissors.[13] For Smith, the utility of these goods mattered less to people than the desire to consume and own. For our purposes, both demand and function are interesting. The very fact that contemporaries were aware of people shelling out on 'trinkets' for bodily care is suggestive of a new vogue for the minutiae of personal grooming.

The relationship between polite conduct and appearance was ambiguous. Vanity and effeminacy were to be avoided at all costs. For men, especially, there was a fine line between sartorial elegance and foppery. Even in the seventeenth century, writers urged young men not to be 'delicately and effeminately appareled', arguing that some types of male dress made them 'strumpet like'.[14] Eighteenth-century 'macaronis', the extreme of male fashion, befuddled commentators even more. Female vanity was no less scrutinised. Poems such as the 1732 'Lady's Dressing Room' lampooned the washes, pastes and ointments of a lady's toilette but also the length of time wasted by women on their appearance. 'Five hours and who can do it less in?/By Haughty Cecilia spent in dressing' mocked the anonymous author.[15] But, while vanity was frowned upon, elegant and harmonious self-presentation was encouraged. Hygiene and cleanliness were also key issues in the maintenance of a healthy, as well as an orderly, body. For John Wesley, the founder of Methodism, cleanliness was 'the mark of politeness'.[16] To be clean, well groomed and cleanly presented was a potential means of distinguishing oneself from the rudimentary sanitary habits of the vulgar.

Personal grooming was not merely a question of refinement, however. Bodies often bore the marks of the society they inhabited. This was an age of rapid industrialisation and urbanisation, the effects of which were easily discernible upon bodily surfaces. A by-product of the huge consumption of coal was the smog that showered promenading beaux and belles in soot and smuts. Georgian towns, although the focus of improvement in the form of wider pavements and broader streets, were still dirty places. An encounter with a passing coach on a wet day bedraggled many a fine set of clothes and begrimed faces and hands. Even the most decorous acts of polite consumption often had a less savoury obverse in the form of wastes and residues. The vogue for snuff begot soiled handkerchiefs and discoloured teeth. The increasing consumption of sugar wrought havoc upon Georgian teeth. Excess hair and face powder, mixed with dried sweat, grew rank over time and encrusted wigs and scalps, in turn

DOI: 10.1057/9781137467485.0008

creating dermatological problems. James Nelson, writing about children but observing bodily maintenance in general, asserted that 'we ought to comb or shave our head, pare our nails, and scour off all the Foulness that Nature throws out upon the surface of the body'.[17] Indeed, a certain amount of intervention was tolerated to maintain cleanliness and form. Whilst debates surrounded the rectitude of cosmetics and their part in disguising the true face, some positively encouraged personal grooming as long as it enhanced natural characteristics, rather than disguising imperfect ones. The physician James MacKittrick Adair emphasised the need to be 'uniformly neat' in all aspects of bodily maintenance since cleanliness and neatness were morally, as well as physically, desirable.[18] Bodily neatness, harmony and elegance were therefore not merely an aesthetic fad. Instead, mastery of the individual body exemplified moral strength and fortitude, which in turn reflected enlightened principles of a mannered and civilised society.[19]

How, though, can the path of this bodily 'ephemera' be mapped? It is hard to track the trajectory of individual instruments; there was no golden age of advertising for ear scrapers as there was for razors in the 1770s and '80s. Tweezers did not achieve fashionable prominence in the same way as steel jewellery, and evidence of actual usage is scarce. This is both frustrating and problematic as it obscures important questions about *how* people attended to their appearance. By exploring other factors from the social importance of bodily areas to the context of instruments for personal grooming, together with the manufacture, design and marketing of the instruments themselves, however, it is possible to chart their growing importance in the daily management of the self. It is to such instruments that this chapter now turns.

Toilette kits and 'necessaires'

Perhaps the ultimate expression of the commoditization of personal grooming was the toilet set or 'necessaire'. Pinpointing the origins of the commercial sale of toilet sets is difficult. They perhaps had their genesis in the French court of Louis XIV. The Sun King popularised the daily *toilette,* performing his morning routines before a retinue of courtiers and guests, transforming both the process and the accoutrements.[20] By the mid-eighteenth century, sets for domestic use were becoming widely available. These could be elaborate and highly decorated items.

DOI: 10.1057/9781137467485.0008

The sheer intricacy and workmanship of some sets strongly suggests a wealthy consumer base, since they were often elaborately decorated, and might be given as gifts to couples setting up home.[21] For ladies, miniature equipage sets, containing everything needed to attend to appearance, from scissors to tweezers, could be worn about the person. Mary Wortley Montague exalted one such set, probably made by the London 'toyman' Charles Mathers, in her poem 'Town Eclogues: Thursday; the Bassette Table':

> Behold this equipage by MATHERS wrought
> With fifty guineas (a great pen'orth !) bought !
> See on the tooth-pick MARS and CUPID strive,
> And both the struggling figures seem to liue.
> Upon the bottom see the Queen's bright face;
> A myrtle foliage round the thimble case;
> JOVE, JOVE himself does on the scissars shine,
> The metal and the workmanship divine.[22]

Toilet sets also found a ready market amongst men. Military and naval officers bought cases of instruments and equipage to facilitate personal grooming in the field. The Grand Tour afforded makers of toilet sets further opportunities to sell small and portable kits for elite travellers to take with them. Typical men's dressing boxes contained soap, powder, oils and scents along with equipment from razors to scissors and curling irons.[23] For some this was a step too far. A correspondent to the *Connoisseur* known only as 'T' complained in 1757 that certain types of 'male beauty [...] aiming to affect the softness and delicacy of the fair sex' not only had dressing rooms 'but a Toilet too'.[24] 'T' went on to describe this gentleman's toilette. Whilst part of the dressing table contained flowers, a mirror and various boxes, a set of 'fantastick equipage' caught his eye, which included toothbrushes, combs, nail nippers and lip salve.[25]

Toilet sets were by no means limited to elites. In fact, they were available from various sellers to suit all pockets. An advertisement by Robert Sangwine of the Strand, London, razor maker, perfumer and 'toyman', illustrates the contents of typical sets. Sangwine specialised in grooming products for both sexes. For men, he sold polished razors, a 'complete set of teeth instruments, fit for Gentlemen to clean their own teeth', a 'great variety of smelling bottles, toothpick cases ... and combs'. For ladies, 'dressing cases of various sizes, that hold powder, pomatum, combes, essences and writing [instruments]'.[26] Customers could make up their own set in cases, pouches and pocketbooks of various shapes

DOI: 10.1057/9781137467485.0008

and sizes, to accommodate their preferred instruments. Another retailer, 'Riccard's Manufactory' in London, was typical of toy 'warehouses' that sold everything from tea tables to dog collars. Amongst the stock in 1770 were instruments like nail nippers, tweezers, 'instruments for cleaning the teeth' and penknives alongside 'medicine chests', etui cases and dressing boxes.[27] Such collections were not only a feature of London retailers. Provincial warehouses also sold similar items in large towns around the country.[28] It is more difficult to find evidence for cheaper sets, but they were clearly available. In 1801, a wooden 'toilette box', valued at only sixpence, was stolen from a London house.[29] Whilst this ad hoc valuation is an unsatisfactory guide to retail value, when compared to the gold 'toilet box' stolen from two jewellers in 1764, and valued at £28, it suggests a more modest item.[30]

As well as large, domestic sets of instruments, various smaller, pocket-sized sets were available, often in elaborate cases known as 'etui' (or 'etwee') cases. These sets contained the most common sorts of grooming instruments for use away from the home, or when quick attention to appearance was required. Trade catalogues offered individuals the chance to peruse a variety of different instruments according to their needs and taste. In many cases catalogues grouped together functional and utilitarian items under collective headings. The 1811 catalogue of Ernst and Co. of London was primarily aimed at gentlemen, but contained sets of useful tools for both sexes. On one page were 'Gentleman's oak chests with turning tools'; another had 'mahogany cases of tools for sportsmen', but there were also pages of items such as nail nippers and tweezers.[31] The ironmongers Ross and Co. in 1797 included nail nippers and ear spoons amongst other useful items, including boot hooks, nutcrackers and lobster crackers.[32] Small, portable instrument sets and cases for travellers contained items such as toothbrushes, toothpicks, nail picks and ear spoons. These could be linked together on a single chain and sold in bespoke cases of leather or fashionable sharkskin (shagreen). Such sets of instruments for the body fitted into a broader market for small functional goods for use in a variety of social situations.

As well as these larger assemblages, however, many people routinely carried small, pocket sets of toilette instruments about them. When John Penny was robbed and murdered by his servant James Hall in 1741, amongst the items stolen were 'a Silver Case for Instruments, covered with Shag-green, a Lancet with a Tortoiseshell Handle, a Pair of Steel Scissars, a Blade of a Knife, a Silver Ear-picker [and] a Pair of Tweezers'.[33]

DOI: 10.1057/9781137467485.0008

When one Elizabeth Jones robbed Mr Berryman of Old Bond Street in 1774, the contents of his pockets included steel scissors, tweezers and a knife.[34] The only contents of a stolen pocketbook belonging to Olivia Harrington in 1777 were a pair of tweezers.[35] In fact, the frequency with which small grooming instruments appear amongst lists of stolen property suggests their relative ubiquity as portable equipment. Attending to appearance was not merely a private matter, and clearly not something done only once a day. Being able to refine appearance in public, or in social gatherings, was important. Small, portable instruments made it possible to carry out running repairs on the fly and avoid potentially embarrassing faux pas caused by errant earwax or food in the teeth. The portability and pragmatic nature of these items are reminders that the management of one's appearance was a process rather than a single event. How, though, were instruments linked to specific body areas and surfaces?

Hands

In 1785, D. Low published his new work, *Chiropodologia*, on the causes of corns and other 'painful or offensive cutaneous excrescences'.[36] Low was one of an emerging group of specialists previously called 'corn-cutters', who began to style themselves 'chiropodists'. Low held shop in London's affluent Berkeley Square, catering to a clientele for whom smoothness of hand and neatness of nail were highly important. Hands were considered strong indicators of an individual's capacity for politeness, as well as a virtual emblem for elegant self-presentation. Something so apparently simple as a well-manicured fingernail spoke of fastidiousness of appearance and also exemplified important factors such as feminine virtue. As Low pointed out, 'a well-shaped hand is no small addition to a well-shaped body...A mere glance at the hand' sufficed for an observer to assess the 'degree of gentility or vulgarity' of a person, or of their 'personal cleanliness or sloth'. Nails that were 'well-formed, well-arranged, transparent, free from spots or furrows... contribute[d] greatly to the beauty of the hand'.[37]

Nicolas Andry had defined ideals of the hand in his seminal *Orthopaedia*. For Andry, hands ought to be 'well-shaped...delicate, pretty long and not square'.[38] Ideally, fingernails should be 'pretty long and of a lively colour, with a small white spot at the root'.[39] Nails that were uneven,

DOI: 10.1057/9781137467485.0008

too large, divided or with their edges 'mangled' were neither desirable nor aesthetically pleasing. There was also a practical element to keeping fingernails in trim. According to Eliza Haywood, servant maids should endeavour to keep their nails 'close-pared' to avoid trapping unsightly bits of meat underneath them and, especially, to evade accusations of sluttishness.[40] As the 'principal organs of touch', hands were vital in the sensory world of the eighteenth century and should be cherished. Legions of recipes in medical self-help books could be found for lotions, pastes and washes to beautify the hands and especially to preserve the whiteness and transparency of the skin.[41] The maintenance of fingernails was therefore of signal importance. If a rough, split or unsightly finger-nail might besmirch a person's appearance and character, attending to them was no small matter. Long fingernails, like beards, could become synonymous with bodily dereliction. They were certainly remarkable. In 1703, Nathaneal Hulme of Bolton was granted relief from parish authorities for having 'nails upon his fingers of a prodigious greatness some of which have been halfe a quarter of a yard longe' – around ten centimetres![42] The author of an account of John Harris, the so-called 'English Hermit' who lived in a Cheshire cave, emphasised the fact that Harris's nails 'had not been cut since he took the life of a hermit ... *which made him appear very frightful* [my emphasis]'.[43] Doubtless they did. But the fact that they were singled out is revealing of a culture in which unkempt fingernails symbolised a body left fallow.

The problem in accessing personal routines like nail care, however, is that people simply had little reason to record what happened to the ends of their fingers unless some accident befell them. It is equally problematic to assume that people across society acted in the same way. For those engaged in daily manual labour, nails were likely continually broken, which kept them short. Nail biting may well have been the usual expedient, but again it leaves no trace in the historical record. Gnawed nails, though, can hardly have met the aesthetic ideal championed by Andry and Low.

Some nail treatments did not involve instruments at all. In the late 1770s, Dr Solomon's Balsamic Corn Extract was one of several similar concoctions that sought to free corn sufferers from the tyranny of the blade. Supposedly endorsed by everyone from the governors of St. Thomas Hospital in London to the wife of the Danish ambassador, 'this balsam cures all kinds of corns without cutting', with the promise that, if the corns returned, so would the sixpence cost of the box.[44]

DOI: 10.1057/9781137467485.0008

Chiropodists also undertook nail cutting, hinting at the potential importance of practitioners in the maintenance of nails. At the end of the advertisement was a further reference to the abilities of 'Dr Solomon' at both dressing corns and cutting nails. One Dr Frankel, newly arrived from Germany in 1792, was 'very famous for cutting nails that grow into the quick of the toes without the least pain or drawing blood'.[45] It seems unlikely, though, not least because of the cost, that practitioners were routinely used for nail cutting.

Prior to the 1750s, the most common instrument for cutting fingernails (and presumably toenails) was a penknife. A sharp blade could be used to incise through dead nail with relative ease and was effective, although difficult to control with precision, especially when holding the knife in the opposite-favoured hand. This method probably involved either paring away at the nail in the manner of sharpening a pencil or instead slicing horizontally through the white (distal) edge of the nail. There is little reason to suppose that the practice was restricted to either sex. It is also possible that family members, friends or servants were called upon to do the cutting. Indeed, individuals wielding sharp blades were a risk to themselves, a fact highlighted by the several deaths attributed to people cutting their own nails. Insufficient care paid to the removal of a nail was considered deadly, especially if the individual cut too close to the quick. In March 1790, one Mr Le Fevre died after 'slightly wounding the tip of one of his fingers through a slip of the knife', the wound becoming putrefied.[46] This was apparently not uncommon. Anthony Henley of Bath died of a mortification of his toe whilst cutting his nails, 'owing to a slip of the knife'. In the opinion of the noted surgeon Samuel Sharp, 'slight wounds at the extremities … often result in terminations'.[47] Other cases reveal the dangers of nonchalance whilst carrying blades. In October 1736, the prominent Dublin physician Dr Vanluen received a 'most dangerous wound in the breast' occasioned by 'his foot slipping as he was walking in his parlour, when he had a penknife in his hands cutting his nails'.[48] As he slipped, the knife plunged into his chest.

Using scissors was an alternative, certainly in popular use in the seventeenth century. A mid-seventeenth-century painting entitled *Old Woman Cutting Her Nails*, reputedly by Nicholas Maes, a pupil of Rembrandt, depicts an elderly subject in deep concentration as she clips a fingernail. Whilst the instrument is not prominent, it is possible to make out the twin blades and handle of a small pair of scissors. Low refers in passing to nails grown so thick 'that no Scissars will cut them'.[49] Scissors were

DOI: 10.1057/9781137467485.0008

commonly sold by 'toymen' from the early eighteenth century, and could be found in toilette sets for various uses. From the second half of the eighteenth century, however, a new option was available in the form of clippers – or 'nail nippers'. One of the first references appears in 1753, when 'nail nippers' were listed amongst the stock of 'Mr Carter's Cane Shop' in Spring-Garden gate, London.[50] Other examples soon followed. By 1785, the advertisements of D. Low included his 'real steel nail nippers', priced at five shillings per pair. The reference to steel in the advertisement again points to its ubiquity and importance both in the construction and aesthetic appeal of devices. It is likely that the tensile springiness of steel was beneficial in these sprung items. Also advertised were his ivory nail models, allowing users to cut their nails to a defined shape 'that they will grow into a beautiful convex'.[51] An unnamed officer of the guards listed nail nippers amongst the 'articles of convenience' requested of his valet in 1779 when travelling to America.[52] While the advertising of nail nippers can hardly be described as a 'campaign', these were, by their nature, small, prosaic and functional items. Unlike razors, which were advertised on their own merits by specialist makers, nail nipper specialists were thin on the ground. As such these, and similar, items were generally subsumed within broader catalogues of instruments or toys. While chiropodists attended to abnormalities and injuries, daily maintenance fell upon the individual, again suggesting a move away from the routine employment of specialist practitioners in bodily care.

Images in trade catalogues reveal small, elegant instruments, mirroring their simple utility. The items in the catalogue of Ross and Co., noted above, were typical of the sorts of styles that could be purchased. Different sizes were available, presumably to suit different hand and finger dimensions, and each was sprung at the handle to enable compact storage with minimal designs around the head of the instrument, along with small handle clasps to keep it closed. Items in other British catalogues reveal very little variation in either design or embellishment. Peter Stubs' catalogue of Lancashire tools contained similar items, but differentiated between sprung and non-sprung nippers, the latter being perhaps easier to control.[53] European collections, however, do contain decorous examples. In the Secq des Tournelles museum in Rouen are several highly decorated items, including intricate and elaborate motifs, which belie their prosaic usage.[54] As a correspondent to the *St James Chronicle* in March 1790 noted, given the potential danger of penknives, and the risk of 'the slightest wound at the extremities of the fingers or

DOI: 10.1057/9781137467485.0008

toes', 'Nail-nippers are therefore the safest instruments to pare nails; – not SHARP KNIVES' (original capitalisation).[55]

Paring the nails, then, was probably an individual pursuit. It required judgement as to when action was necessary, and also decisions about the amount of nail to be removed. How far aesthetics intruded into the thoughts of users across the social spectrum is difficult to fathom. Nonetheless, the numbers and variety of instruments available, together with the apparent importance attached to maintaining the nails, especially for the upper echelons of society, suggests a basic routine that was invested with new meaning and importance in the eighteenth century.

Face .

If any part of the eighteenth-century body was central to the conveyance of politeness, it was the face. Expression, physiognomy and symmetry all augured personality. What was writ upon the face reflected the character that lay underneath, so managing facial appearance to give the best impression was important. Little attention has been paid to the management of the face and the role of technologies in constructing the facial image. As discussed in Chapter 2, razors were key instruments in maintaining men's faces. But, in fact, general depilation of the face was an issue for both sexes. Tweezers were important in this process.

Eyebrows, for example, were important barometers of character, 'placed by nature as an ornament to the face'.[56] The popular pseudomedical sex manual *Aristotle's Masterpiece* noted that '[f]or some men, crimes are in their forehead writ'.[57] Arched eyebrows in both sexes suggested a proud, high-spirited and bold individual. A person whose eyebrows bent downwards was likely a 'penurious wretch' and 'full of malice in his heart'. Thick eyebrows suggested clownishness and bespoke an unlearned mind.[58] Andry also commented upon eyebrow 'styles'. According to Andry, 'the space between the two Eye-Brows ought to be quite bare'; having eyebrows that joined in the middle was a deformity, giving the face 'an unlucky look'.[59] Eyebrows that only met whilst the 'brow was contracted', in other words when a person frowned, suggested instead a 'thoughtful and melancholy disposition'.[60] It was vital for Andry that eyebrows pointed the correct way, from the nose to the temples. A parent whose child's eyebrows pointed the opposite way 'cannot too quickly set about removing this deformity'. This could be achieved by gentle brushing

DOI: 10.1057/9781137467485.0008

with a small brush 'such as people rub the teeth with'.[61] But while Andry disliked arched eyebrows, for others they represented the very acme of beauty. Along with a clear complexion, bright eyes and handsome nose they were commonly regarded as key indicators of beauty.[62]

In general, excessive facial hair was viewed negatively. For women, facial hair suggested bodily decline and the termination of sexual activity. According to a 1733 nursing text, 'most old women have their chins cover'd with Whitish hair'.[63] Various compounds were available to help ladies remove unsightly hair discreetly. 'Mr Gibson's Curious Compound' promised to 'take off HAIR growing on ladies cheeks ... which must be owned to be a great blemish to the fair sex' and also 'thins large eyebrows and turns them into an arch'.[64] For Monsieur Mosenau, lately arrived in London in 1778, 'superfluous hairs' on the face of a lady were nothing less than a deformity, requiring the careful deployment of 'toilet arts'![65]

Facial hair held multiple meanings. People were prepared to spend time and money on products to control it. Cosmetics and remedies were one option. Instruments, however, were also a logical choice. Scissors offered one means of removing excessive eyebrow hair. Andry recommended scissors as best for thinning or cutting overlong eyebrows – 'there being no other Expedient that can be of service in this case'.[66] Also popular were tweezers. These were common across many trades, from cloth making to glass blowing, but extremely useful in plucking hair from eyebrows and chins. The use of tweezers by men in plucking out beard hairs was remarked upon as a characteristic of non-European cultures. Chinese men were said to 'pluck the hair from their cheeks with tweezers.[67] In 1789, traders amongst the American Indians found 'tweezers for beard-plucking a very profitable article of commerce'.[68] The practice of beard plucking, rather than shaving, was also linked to barbers. In 1745, reporting on his voyage to Holland, 'CW' remarked on barbers using razors, tweezers and perfumes on the faces of Dutch beaus.[69]

For women, tweezers were important toilette instruments. They were certainly valuable enough to warrant a specific entry in the will of the Duchess of Shrewsbury who, in 1726, left her tweezers, along with her snuffbox and some clothes, to her servant.[70] It is interesting that, in a bequest entry totalling £200, tweezers should warrant an individual entry. References in popular culture further suggest the ubiquity of the item as an essential tool to be carried about the person. In the poem 'The Nightingale a Tale', which appeared in British newspapers in the 1720s, an officious mother was said to always be close to her daughter 'just like

DOI: 10.1057/9781137467485.0008

her tweezers by her side'. Men too were clearly consumers and users of tweezers. In 1733, the 'Ingenious' Edward Pinchbeck sold many 'curious toys' from his Fleet Street shop. His list included 'tweezers for men and women', suggesting both differences in design and also, importantly, that men were using them for depilation.[71]

Many different types of tweezers were available for all pockets and purposes. In 1745, one Cornhill goldsmith advertised gold and metal examples.[72] Amongst the possessions of the bankrupt Bishopsgate jeweller Francis Cooper in 1744 were 'Gold tweezers and equipage richly chas'd'.[73] These proved highly desirable items for thieves. In 1739, several ladies attending the Princess of Wales at Vauxhall Gardens fell victim to a pickpocket. Whilst one lady lost her purse, another was reportedly relieved of her tweezers.[74] Tweezers could be found for sale across various trades, from jewellers to razor makers. Examples listed in ironmongers' catalogues depict various sorts but, interestingly, nearly all include ear spoons (for removing ear wax) on the handle of the instrument.[75] Why ear spoons in particular, rather than, say, toothpicks, is unclear. Nevertheless, the frequency does suggest that the combination, and associated practices, were fairly common.

The mouth

As Colin Jones has highlighted, the last decades of the eighteenth century brought great changes in attitudes towards cleaning and maintaining teeth. In France, as Jones argues, something like a 'smile revolution' occurred when, for a relatively brief period of time, a toothy smile became de rigueur. Before 1780, to show the teeth when smiling was uncouth and unbefitting.[76] In the decade after, however, smiling was the very height of polite fashion, prompting a slew of portraits of beaming subjects.[77] The growing emphasis upon dental appearance occurred at a time when dentistry emerged as a profession in France. After the 1750s, the humble tooth drawer was sacrificed on the altar of the professional *dentiste* as oral care became the dominion of the expert.[78] As with many aspects of French enlightened culture, this new vogue for dental care was not long in crossing the English Channel.[79] There is insufficient space here to explore the extent of any similar smile 'revolution' in England. However, a trend for dental technologies in Britain in this period is apparent.

DOI: 10.1057/9781137467485.0008

As in France, for example, care of teeth was a central part of bodily maintenance in Britain in the later eighteenth century. Indeed, English authors were at pains to stress the impropriety of bad teeth, whilst also recognising that the English lagged some way behind their French neighbours. For women in particular, gappy teeth were a considerable social impediment. As the author of one ladies' conduct book put it in 1791, tooth loss was an injury to beauty, causing the cheeks to 'fall in and look lanky' and the 'whole visage to appear no larger than that of a child'.[80] The surgeon William Bennett (literally) cited teeth at the literal forefront of health and beauty. 'The teeth are often the first remarked in describing an elegant figure which everyone naturally wishes to possess'.[81] For Bennett, dental care and, indeed, the desire for elegance, was 'natural'. To neglect teeth was to ignore their 'vast importance'. It therefore 'behove[d] everyone to perform this office unremittingly'.[82] Nonetheless, the poor state of British teeth was widely remarked upon. Paul Juillon, the author of a dental treatise, denounced the parlous state of British teeth among all but the denizens of the 'Fashionable World'.[83] His barbed point was that people of fashion recognised the importance of good a set of teeth and were prepared to act in order to get them.

Poor oral health could indeed be socially inhibiting. As Stephen Greenblatt argued, managing and controlling the body's waste products was an essential route to civility and, along with the mastery of appropriate manners and rituals, a key marker of status.[84] Whilst sweat, saliva and faeces are obvious examples, dental tartar was also part of this process. Aside from the fact that tartar was considered a contributory factor in conditions from fevers to heartburn, bad breath also inhibited politeness.[85] The 'sanious effluvia' and 'putrid exhalations' of those with rotten teeth diminished a person's character, 'rendering them odious to those in whose company they should fall'.[86] Worse still, the spittings and stutterings of a toothless speaker invited mockery, making a 'person of real learning' a figure of fun.[87] Sufferers might choose to visit one of the many tooth specialists becoming popular across Britain. Men like Robert Law, the 'Ingenious Mechanick' of Birmingham, 'cleanseth the teeth, taking away all their tartrous Scales, or flimsy or muddy humours'.[88] In 1776, Northampton was visited by the itinerant 'Mr Crawcour, Dentist' offering services from cleaning and scaling to stump removal, and solemnly promising to visit the town at least once every year.[89] One Mr Moor 'attended all various operations of the Teeth and Gums' in Reading and Oxford in 1779, even claiming to supply artificial

DOI: 10.1057/9781137467485.0008

gums.[90] Precisely how these would be attached to the palette is unfortunately not recorded.

Dentistry was expected to begin at home, and instruments were key to the process. A raft of new publications sought to educate the public on the proper care of their teeth. Toothpicks were one means of preserving both the appearance and health of the teeth. 'Surgeon-Dentist' and Royal Society member Michael Le Maitre advised readers 'after every meal to pick out every bit of flesh that might be sticking out between the teeth'.[91] Food waste left in the mouth invited putrefaction. Others pointed out the dangers of small particles of cheese, which, if not removed, could swiftly bring on caries.[92] Daily use of a toothpick to remove embedded food matter was therefore encouraged. One surgeon dentist even suggested an early version of dental floss, using horsehair or other thread between two toothpicks to get between teeth.[93]

As with other instruments, toothpicks were available for all budgets. Plain metal examples cost a few pennies from toy sellers and perfumers like James Love of Haymarket, while expensive examples made from precious and 'chased metal[s]' again proved tempting for pickpockets.[94] Even for something so prosaic as the toothpick, early forms of branding were emerging. By 1770, 'The Real Lisbon Toothpicks' were advertised for the nobility and gentry everywhere from Oxford to Dublin.[95] A selection of 'New-invented toothpicks from the Nuns of Santa Rosa', Portugal, could be had from Thomas Paynes, bookseller in Bishopgate Street.[96] Elaborate cases were often made of modish substances like ivory and tortoiseshell, and inlaid with gold, silver or shagreen.[97] In fact, these were highly desirable in their own right and popular items to request from London-based family or acquaintances. Writing to her friend the Baroness Grantham, Countess Annabel, Lady Lucas entreated the Baroness to purchase a toothpick case for her. Only one from a London maker would do.[98] Gold and silver cases were the very height of conspicuous luxury. Picking the right one was seemingly a matter of great import.[99] The character Robert Ferrars in Jane Austen's *Sense and Sensibility* is first noticed by the Dashwood ladies whilst he is in a shop 'giving orders for a toothpick-case for himself, and till its size, shape, and ornaments were determined, all of which, after examining and debating for a quarter of an hour over every toothpick-case in the shop, were finally arranged by his own inventive fancy'.[100] A fictional depiction yes, but nonetheless interesting that, of all items, a toothpick case should be selected. This was perhaps a device to emphasise the snobbery and

DOI: 10.1057/9781137467485.0008

arrogance of the character in regarding it as a 'necessity of life', but it marks out its ready status as a desirable item of toilette. Elaborate cases also highlight the public dimension to personal grooming. As with other forms of bodily maintenance, it was acceptable to pick the teeth in public, albeit with reasonable discretion. French ladies reputedly hid their mouths whilst using toothpicks, 'through excess of delicacy'.[101] Tongue scrapers were another useful, but seldom advertised, item. John Cleland's *1761 Institutes of Health* contained an entire chapter on oral care. The mouth was to be kept 'religiously clean' and the tongue scraped with a whalebone scraper, after which the mouth should be rinsed with fair water.[102] A 'common quill pick-tooth' was to be used to remove unsightly particles of food.[103]

Toothpicks enter the historical record in a variety of circumstances, which itself suggests their relative ubiquity. Silver, metal and even wooden toothpicks, along with cases made from everything from paper to precious metals, for example, were commonly listed amongst stolen goods. These ranged in value from a few pence to several pounds for an ornamental case. Other passing references suggest that carrying toothpicks was routine. In August 1775, one Lieutenant Colonel Abercromby was killed when, leading a charge, he received a shot to his groin of such force that a toothpick case in his breeches pocket was pushed inside his body.[104] The fact that an officer carried a toothpick case about his person is perhaps not surprising; that he took it into battle is more so. In popular culture the practice of picking teeth was even used metaphorically to suggest a routine, nonchalant or even relaxing activity. Charles Churchill's 1761 epistle to Robert Lloyd suggested thoughts that rambled 'in a tittle-tattle tooth-pick way'.[105] The main character in Coriat Junior's *Another Traveller* longed for 'uninterrupted, listless, tooth-pick ease'.[106] The suggestion overall is that picking the teeth was a common enough practice, and not necessarily just by elites.

Some, however, cautioned against overly harsh treatments. Whilst scaling was recommended, overzealousness by either practitioner or individual, particularly in the use of metal scrapers and instruments, risked enamel damage and tooth loss.[107] In 1754, Lord Chesterfield cautioned his son to avoid the 'picks, irons etc' which had totally destroyed his own teeth, 'so I have not now above 6 or 7 left'.[108] Paul Juillon noted the damage caused by metal toothpicks, promoting his own 'toothpick brushes' to remove food particles between the teeth.[109] In fact, by the 1780s, toothbrushes were the coming thing, and considered a safe means

DOI: 10.1057/9781137467485.0008

to gently remove thickly encrusted tartar and restore the natural white-ness of the teeth. Dental practitioners like Hugh Moises keenly promoted their use, but stressed that only soft-bristled brushes were safe to use.[110] Toothbrushes could be purchased from specialist retailers, like Trotters of Christchurch, patentees of the 'Asiatic Tooth Powder' from whom the 'India Tooth Brush' could be bought for a shilling.[111]

In fact, unlike several other small instruments of the body, tooth-brushes were sufficiently prominent as to warrant individual advertise-ments. Interestingly, makers of other body products often sold them. For a whole year between 1796 and 1797, the London razor maker Benjamin Kingsbury published a series of puffs for his 'tooth-brushes of superior excellence'.[112] As with his razors, Kingsbury stressed the workmanship in his toothbrushes, claiming they lasted longer and were less prone to looseness in the bristles than those of his competitors. In another exam-ple of brand assurance, each of Kingsbury's brushes was stamped with his name. By September 1797, the advertisement occupied a whole column of text and included pouches to hold tooth powder and brushes togeth-er.[113] In the provinces, manufacturers sold other branded goods such as 'Bott's tooth powder and brushes' alongside other dental products such as pastes and dentifrices.[114] As items of personal hygiene, toothpicks and toothbrushes were highly esteemed, and retailers were therefore keen to highlight the range, prices and styles of the goods they sold.[115]

For those whose teeth had already deserted them, a last resort was to take advantage of the many artificial teeth and other contrivances introduced to shore up depleted or neglected mouths. Francis Spilsbury suggested that it was common practice for ladies to press white wax into the gap, then cutting it into shape with a knife, which could last up to four days.[116]

Conclusion

Instruments of various kinds helped people maintain, or aspire to, expectations of bodily neatness, harmony and elegance. Elsewhere in this book it is shown that new steel technology brought material improvements to the design or function of particular goods, but in turn acted as a vector for the conveyance of polite appearance. In this sense, steel objects carried a deeper meaning beyond their basic utility. In this chapter, however, the interplay between object and practice is subtler. It

DOI: 10.1057/9781137467485.0008

was not necessarily that the objects themselves were materially altered. While metallurgical technologies doubtless made small refinements possible, they did not transform the effectiveness of objects like nail nippers, tweezers or toothpicks. Instead, a new focus upon the minutiae of appearance, and upon individual elements of the body, invested personal grooming, and the instruments and practices it encompassed, with new meaning.

Whilst nail cutting had always been a necessary expedient, new aesthetic and social ideas relating to hands rendered nail care an important part of grooming rituals. Plucking eyebrows, together with other forms of depilation, chimed with ideals of self-mastery and the careful management of the public face. A new focus upon teeth, and the deleterious effects upon both health and appearance of dental caries, made toothbrushes and toothpicks desirable items of equipage. In each case, small, prosaic instruments were subsumed into polite consumption. While specialist practitioners, from dentists to corn cutters existed, people took responsibility for their own grooming practices, and found a ready market of goods to serve them. Although in aggregate it is difficult to find particular 'booms' for individual goods, instruments for the body were appearing in advertisements more frequently, in more ornate forms, and also across a broader price range. Some items, such as gold toothpicks, reflected luxury beyond utility. Toilet sets could be purchased in very basic versions but also as ornate and expensive 'trophy' items for travelling gentlemen. This willingness to spend on expensive versions of very basic instruments also suggests a practice shifting from the private to the public sphere. People invested in luxurious cases for their instruments or hung them prominently on chatelaines. Attending to one's appearance in public was acceptable, perhaps even essential, to ensure that an errant body did not confound social expectations.

Rather, therefore, than being mere 'toys', or 'trinkets of frivolous utility', small instruments were part of a continuum of bodily maintenance in which everything from tweezers to toothpicks contributed to enlightened self-fashioning and the articulation of a socially pleasing body.

Notes

1 Quoted in Virginia Smith, *Clean: A History of Personal Hygiene and Purity* (Oxford: OUP, 2007), 225.

DOI: 10.1057/9781137467485.0008

2 John Styles, *The Dress of the People: Everyday Fashions in Eighteenth-Century England* (New Haven and London: Yale University Press, 2010), 3.

3 Neil McKendrick, 'The Consumer Revolution of Eighteenth-Century England' in Neil McKendrick, John Brewer and J.H. Plumb (eds), *The Birth of a Consumer Society* (Bloomington: Indiana University Press, 1982), 9–10.

4 For examples see Maxine Berg and Helen Clifford (eds), *Consumers and Luxury* (Manchester: Manchester University Press, 1999); Maxine Berg, *Luxury and Pleasure in Eighteenth-Century Britain* (Oxford: Oxford University Press2005); Amanda Vickery, *Behind Closed Doors: At Home in Georgian England* (New Haven and London: Yale University Press, 2009).

5 Smith, *Clean*; David M. Turner, 'The Body Beautiful' in Carole Reeves (ed.), *A Cultural History of the Human Body in the Enlightenment* (London: Bloomsbury, 2010), 130; See also Lynn Festa, 'Cosmetic Differences: The Changing Faces of England and France', *Studies in Eighteenth-Century Culture* 34 (2005), 25–54; Richard Corson, *Fashions in Makeup* (London: Peter Owen, 1972).

6 Morag Martin, 'Doctoring Beauty: The Medical Control of Women's *Toilettes* in France, 1750–1820', *Medical History* 49:3 (2005), 351–68.

7 Ibid., 353.

8 Virginia Smith, 'Cleanliness: Ideas and Practice in Britain, c. 1770–1850', unpublished PhD diss., London School of Economics, 1985), 133–5. See also Kathleen Brown, *Foul Bodies: Cleanliness in Early America* (New Haven, CT: Yale University Press, 2009).

9 Liliane Hilaire-Pérez and Christelle Rabier, 'Self Machinery? Steel Trusses and the Management of Ruptures in Eighteenth Century Europe', *Technology and Culture*, 54:3 (2013), 461–2.

10 A process that Virginia Smith views as being complete by 1850; Smith, 'Cleanliness', 193.

11 Turner, 'The Body Beautiful', 126.

12 Maxine Berg, 'New Commodities, Luxuries and Their Consumers in Eighteenth-Century England' in Berg and Clifford (eds), *Consumers and Luxury*, 69.

13 D.D. Raphael and A.L. Macfie (eds), Adam Smith, *The Theory of Moral Sentiments* (Oxford: Oxford University Press, 1976), 181. See also Jonathan Lamb, *The Evolution of Sympathy in the Long Eighteenth Century* (London: Pickering and Chatto, 2009), 54.

14 Quoted in Alexandra Shepherd, *Meanings of Manhood in Early Modern England* (Oxford: OUP, 2008), 29.

15 *Read's Weekly Journal or British Gazetteer*, Saturday, 24 June 1732.

16 Quoted in Virginia Smith, *Clean: A History of Personal Hygiene and Purity* (Oxford: OUP, 2007), 226.

17 James Nelson, *An Essay on the Government of Children* (London: Printed for R. and J. Dodsley, 1756), 139.

DOI: 10.1057/9781137467485.0008

18   Smith, 'Cleanliness', 148.

19   Ibid.

20   Smith, *Clean*, 228.

21   Ibid.

22   Alexander Pope, *Court Poems* (Dublin: Reprinted by S. Powell, 1716), 5. Although Pope is given as the author, the poem has been attributed to Mary Wortley-Montague.

23   Richard Corson, *Fashions in Makeup*, 237.

24   George Colman, *The Connoisseur by Mr. Town, Critic and Censor General ... Volume 2* (London: Printed for R. Baldwin, 1757), 230.

25   Ibid.

26   Advertisement, 'Shaving Cases', *Gazetteer and New Daily Advertiser* (London, England), Tuesday, 1 September 1778

27   Advertisement for 'Riccard's Manufactory, London', University of Oxford, Bodleian Library, John Johnson Collection of Printed Ephemera, Advertisements, no. 1, available at http://ojohnjohnson.chadwyck.co.uk.lib. exeter.ac.uk/search/results.do, accessed 16 June 2014.

28   For examples, see Richard Croker's advertisement in *The Reading Mercury and Oxford Gazette, etc*, Monday, 26 April 1773; pg. 1; J. Newland of Stamford, Lincolnshire, *The Lincoln, Rutland and Stamford Mercury* Friday, 4 March 1796; pg. 1 and Mr Craven of Leeds – *The Leeds Intelligencer*, Monday, 20 May 1805; pg. 1.

29   Old Bailey Proceedings Online, hereafter OBPO, 14 January 1801, trial of ANN SAUNDERS, available at http://www.oldbaileyonline.org/browse. jsp?id=t18010114–65-defend684&div=t18010114–65#highlight, accessed 15 September 2014.

30   OBPO, 22 February 1764, trial of John Taylor Lot Lash, http://www. oldbaileyonline.org/browse.jsp?ref=t17640222–10 accessed 15 September 2014.

31   Victoria and Albert Museum MS E126/96, Trade catalogue of Ernst and Co. Ltd, 1811, 3, 4, 11, 18, 22.

32   Victoria and Albert Museum MS E10–98 and E36–98, Trade catalogue of Ross and Co. Ironmongers, c. 1797, ff. 8, 11, 27.

33   OBPO, Ordinary of Newgate's Account, 14 September 1741, http://www. oldbaileyonline.org/browse.jsp?ref=OA17410914 accessed 15 September 2014.

34   OBPO, 3 June 1772, trial of ELIZABETH JONES, http://www.oldbaileyonline. org/browse.jsp?ref=t17720603–14, accessed 15 September 2014.

35   OBPO, 9 April 1777, trial of PIERCE DONNOVAN WILLIAM KERWIN, available at http://www.oldbaileyonline.org/browse.jsp?ref=t17770409–14, accessed 15 September 2014.

36   D. Low, *Chiropodologia, or A Scientific Enquiry into the Causes of Corns, Warts, Onions and Other Painful or Offensive Cutaneous Excrescences ...* (London: Printed by J. Rozea, 1785).

DOI: 10.1057/9781137467485.0008

37 Ibid., 108–9.
38 Nicholas Andry, *Orthopaedia or the Art of Correcting and Preventing Deformities in Children* (London: 1743), 152.
39 Ibid., 153.
40 Eliza Haywood, *A Present for a Servant-Maid. Or the Sure Means of Gaining Love And Esteem* (London: printed by and for George Faulkner, 1744), 11.
41 Ibid., 153, 156; for examples of recipes, see Amelia Chambers, *The Ladies Best Companion...* (London: 1775), 153–4; Pierre-Joseph Bu'choz, *The Toilet of Flora...* (London: 1772), 57, 82, 157.
42 Lancashire Archives, MS QSP/905/7, Quarter sessions records 1703/4.
43 Anon, *A Full Account of Mr John Harris, the English Hermit* (Banbury: Printed by T. Cheney, 1800).
44 Advertisement, 'Dr Solomon's Balsamic Corn Extract', *Morning Post and Daily Advertiser* (London, England), Friday, 22 October 1779.
45 Advertisement, 'Just Arrived from High Germany', *Morning Chronicle* (London, England), Monday, 29 October 1792.
46 'Dangers of Wounding the Extremities', *St. James's Chronicle or the British Evening Post* (London, England), 11 March 1790–13 March 1790.
47 Ibid.
48 'News', *London Daily Post and General Advertiser*, Thursday, 28 October 1736.
49 Low, *Chiropodologia*, 118.
50 Advertisement, 'Carter's Cane Shop', *Public Advertiser*, Wednesday, 2 May 1753.
51 Advertisement, 'Corns and Nails', *Morning Post and Daily Advertiser*, Friday, 18 February 1785.
52 'Intelligence a la Militaire', *General Advertiser and Morning Intelligencer*, Thursday, 8 April 1779.
53 Manchester Archives MS L24/1 (Box 24), 'The Prices of Lancashire Tools &c Manufactured by Peter Stubs, Warrington', undated, late eighteenth century, 16.
54 Images in Henry René D'Allemagne, *Decorative Antique Ironwork: A pictorial treasury* (New York: Dover Publications, 1968), 237.
55 Anon, 'Dangers of Wounding the Extremities', *St James Chronicle or the British Evening Post*, 11–13 March 1790.
56 Anon, *Aristotle's New Book Of Problems, Set Forth by Way of Question and Answer. To Which Are Added, a Great Number from Other Famous Philosophers* (London: 1725), 34.
57 Anon, *Aristotle's Compleat Master Piece. In Three Parts; Displaying the Secrets of Nature in the Generation of Man* (London: 1749), 102.
58 Ibid., 102.
59 Andry, *Orthopaedia*, 32.
60 Ibid., 34.

DOI: 10.1057/9781137467485.0008

61   Ibid., 35.

62   These characteristics are commonly referred to in popular novels. For examples, see Anon, *The History of Miss Harriott Fitzroy and Miss Emilia Spencer* (London: printed by W. Hoggard, 1767), 239; Anon, *Eliza: Or the History of Miss Granville* (London: printed for W. Hoggard, 1766), 172; Anon, *Beatrice, or the Inconstant… Volume 1*(London: printed for William Lane, 1788), 111.

63   Anon, *The Art of Nursing: or the Method of Bringing up Young Children According to the Rules of Physick for the Preservation of Health and Prolonging Life* (London: Printed for John Brotherton, 1733), 84.

64   Advertisement, 'To Take Off Hair Growing on Ladies Cheeks', *Lloyd's Evening Post*, 8–10 August 8 1764.

65   Advertisement, 'To the Fair Sex', *Morning Post and Daily Advertiser*, 9 July 1778.

66   Andry, *Orthapaedia*, 36.

67   Anon, 'To the Printer of the London Chronicle', *London Chronicle*, 10–12 August 1773.

68   Anon, 'To the Printer of the London Chronicle', *London Chronicle*, 26–28 November 1789.

69   CW, 'A Voyage to Holland, a Description of a Storm, the Dutch Greater Beaux than at Whites, and More Gaudy Than the French', *Penny London Post or The Morning Advertiser*, 6–8 March 1745.

70   *Weekly Journal or British Gazeteer*, 9 July 1726.

71   Advertisement, 'To Prevent for the Future…' *The Daily Post*, 23 March 1733.

72   Advertisement, *Daily Advertiser*, Thursday, 11 July 1745.

73   'To Be Sold by Auction by John Ashley', *General Advertiser*, 23 May 1747.

74   'London', *Daily Post*, 24 May 1739.

75   Victoria and Albert Museum MS E10/98-E36–98, page from Ross & Co. Catalogue, c. 1797, f.11; see also Victoria and Albert Museum MS E126/96, Trade Catalogue of Ernst and Co., c. 1810, ff. 11,18,22.

76   Colin Jones, *The Smile Revolution in Eighteenth Century Paris* (Oxford: OUP, 2014), 9–10.

77   Ibid., 55–7.

78   Ibid., 81.

79   Virginia Smith, *Clean: A History of Personal Hygiene and Purity* (Oxford: Oxford University Press, 2007), 230–1.

80   Anon, *The Ladies Library or Encyclopaedia of Female Knowledge* (London: 1790), 434.

81   William Bennett, *A Dissertation on the Teeth and Gums, and the Several Disorders to Which They Are Liable* (London: 1779), 3.

82   Ibid., 4.

DOI: 10.1057/9781137467485.0008

83 Michael Le Maitre, *Advice on the Teeth: with Some Observations and Remarks* (London: 1782), ix. See also Ruspini's comment on British teeth, quoted in Anne Hargreaves, 'Dentistry in the British Isles' in Christine Hillam (ed.), *Dental Practice in Europe at the End of the Eighteenth Century* (Amsterdam: Rodopi, 2003), 195.

84 Stephen Greenblatt, 'Filthy Rites', *Daedalus*111:3 (1982), 2.

85 Paul Eurialius Juillon, *A Practical Essay on the Human Teeth* ... (London: Printed for the author, 1781), 9.

86 Bennett, *Dissertation,* 6.

87 Ibid., 4.

88 Advertisement, 'This Is to Give Notice ...', *The Birmingham Gazette or the General Correspondent,* 16 November 1741, 4.

89 Advertisement, 'Mr Crawcour, Dentist at the George Inn', *Northampton Mercury,* 25 November 1776, 2.

90 Advertisement, 'Mr Moor, Dentist from Oxford', *The Reading Mercury and Oxford Gazette,* 12 April 1779, 1.

91 LeMaitre, *Advice on the Teeth,* 65.

92 Richard Curtis, *A Treatise on the Structure and Formation of the Teeth* ... (London: 1769), 23–4.

93 Robert Wooffendale, *Practical Observations on the Human Teeth by Robert Wooffendale, Surgeon-Dentist, Liverpool* (London: 1783), 153.

94 James Love, *All Sorts of Italian, French and English Perfumes and Powders, with a Variety of Choice and Curious Articles* (London: 1800); Advertisement, 'Found in the Custody of a Suspicious Person', *Gazetteer and New Daily Advertiser,* 1 September 1778.

95 Advertisement, 'Alice, Savage and Kavanagh', *Dublin Mercury,* 28 January 1769; Advertisement, 'Richard Warren, Perfumer in Marylebone', *London Evening Post,* 8 August 1771.

96 Advertisement, 'To the Virtuous in Flowers &c', *London Daily Advertiser,* 13 February 1753.

97 For example, see 'Great Choice of Mahogony Shaving Cases', *St James Chronicle or the British Evening Post,* 5 August 1779; 'Ladies Dressing Cases', *General Advertiser,* 2 May 1786.

98 Bedfordshire and Luton archives, MS L30/13/12/12, Letter from Annabel, Lady Lucas to Mary Jemima Robinson, exact date unknown, 1774.

99 Abraham Langford, *A Catalogue of the Genuine Stock in Trade of Mr Stephen Quillet, Jeweller and Goldsmith* (London: 1751).

100 Quoted in Emily Auerbach, 'An Excellent Heart: Sense and Sensibility' in Harold Bloom, (ed.) *Jane Austen* (London: Chelsea House, 2009), 264.

101 Anon, 'Observations on the Manners and Customs of the French', *Oxford Magazine or University Museum,* 8, (1772), 172. My thanks to Sarah Murden for this reference.

DOI: 10.1057/9781137467485.0008

102   John Cleland, *Institutes of Health* (London: Printed for T. Becket, 1761), 1–2.

103   Ibid., 2.

104   Anon, *The Scots Magazine*, 37 (1775), 463.

105   Charles Churchill, *Night, An Epistle to Robert Lloyd* (London: Printed for the author, 1761), 6.

106   Coriat Junior, *Another Traveller or Tritical Observations made on a Journey Through the Netherlands* (London: printed for Johnson and Payne, 1766), 168.

107   Bennett, *Dissertation*, 50.

108   Lord Chesterfield, Letter CXCVI, London, 15 February 1754, available at http://www.gutenberg.org/files/3361/3361-h/3361-h.htm, accessed on 25 November 2014.

109   Juillon, *Practical Essay*, 72.

110   Hugh Moises, *An Appendage to the Toilet: Or an Essay on the Management of the Teeth. Dedicated to the Ladies* (London: 1798), 22.

111   Advertisement, 'Trotter's Asiatic Tooth Powder', *World*, Thursday, 2 June 1791.

112   Advertisement, 'Tooth-Brushes of Superior Excellence', *Courier and Evening Gazette*, 24 October 1796.

113   Ibid.

114   For example, see Advertisement, 'Parker's Toy Shop in Derby', *The Derby Mercury*, 30 January 1800.

115   A point made by Nancy Cox and Karen Dannehl – 'Tooth brush – Toy card', *Dictionary of Traded Goods and Commodities, 1550–1820* (2007). URL: http://www.british-history.ac.uk/report.aspx?compid=58898, accessed 25 November 2014.

116   Francis Spilsbury, *Every Lady and Gentleman Their Own Dentist, as Far as the Operations Will Allow…* (London: 1791), 55.

DOI: 10.1057/9781137467485.0008

4
New Ways of Seeing: Sight, Spectacles and Self-Fashioning

Abstract: *The period between 1700 and 1850 was transformative in the manufacture, consumption and conception of eyewear. Whilst scientific instruments have garnered much attention from historians, spectacles, which are effectively optical instruments for the body, have not. The introduction of cast steel transformed spectacles from objects balanced on the end of the nose, to sprung-armed items worn on the head. Some of this demand was driven by changing fashions, such as wig wearing, with new spectacles designed to fit around the wig. They also became more decorous and, rather than being hidden away, more 'public'. New enlightened emphases upon vision saw eyes revered as principal organs of sense. Withey's groundbreaking study charts the social, cultural and technological changes, from literacy to coffee-house culture, which saw spectacles shake off earlier associations with deficiency, and begin to represent learning and sagacity.*

Withey, Alun. *Technology, Self-Fashioning and Politeness in Eighteenth-Century Britain: Refined Bodies.* Basingstoke: Palgrave Macmillan, 2016. DOI: 10.1057/9781137467485.0009.

In 1750, James Ayscough of London was one of several opticians taking advantage of cheap newspaper advertising space to promote their wares. From his shop at the sign of the Great Golden Spectacles in Ludgate Street, Ayscough made and sold his 'Superfine Crown Glass' eyeglasses in various colours and mounts.[1] These, he argued, were far superior to those 'sold by wholesale in the country, or hawk' d by peddling Jews about the streets'.[2] Over the course of the eighteenth century, the form and function of spectacles changed. Metallurgical innovations allowed changes to the material components of spectacle manufacture, transforming them from the traditional form, resting on the wearer's nose, to a new design with arms to adhere to the head. The appearance of spectacles also changed. Whilst traditional materials like horn, tortoiseshell and leather remained, the range expanded to include newly modish metals like steel and silver. While metal spectacles had existed since the sixteenth century, new materials were emphasised, and there was a continual process of innovation in design and manufacture. Accompanying these changes were important shifts in attitudes towards spectacles. Indeed, the period between 1650 and 1850 was transformative in terms of the manufacture, consumption and understanding of eyewear. Spectacles came to symbolise a new fascination with sight, both physically and as a metaphor for intellectual enquiry. This period, longer in duration to others in this book, also saw changes in the function of spectacles. After the 1720s, spectacles changed from exogenous items, intended to be held up to the eyes, to items that were to be 'worn'. As such they became subsumed within the wider context of appearance. Rather than simply being workaday, utilitarian objects, spectacles became enmeshed in a complex web of meanings encompassing new technologies, self-fashioning and ideas about the body, but also a series of new environments to which the body, and particularly the eye, was becoming exposed.

Aside from the landmark studies by Corson and, more recently, Rosenthal, little attention has focussed on the cultural dimension of eyewear, or the place of optical aids in broader changes to concepts of the body.[3] Histories of individual firms and makers such as Dollond and Aitchison have tended to be descriptive and pictorial rather than analytical, with a focus upon design.[4] In the 1920s, Thomas Court and Moritz Von Rohr provided useful overviews of the technological innovation of opticians and the general development of spectacle technology.[5] Spectacles occasionally intrude into studies of telescopes and lenses,

DOI: 10.1057/9781137467485.0009

while a raft of popular texts serves the huge market for the collection of antique spectacles.[6]

Most commonly they are subsumed within the historiography of the enlightenment trade in scientific instruments. Spectacles were homologous with the trade of optical instrument makers, one of a triumvirate of mathematical, optical and philosophical instruments. General instrument studies by Alison Morrison-Low and others have recovered the role of popular science, instrument collection and public demonstration, in fixing science within the public consciousness.[7] Recent studies of prominent makers reflect a rising interest in instrument technology and retail.[8] But less attention has focussed upon spectacles or, more specifically, their wearers in the eighteenth century. Less complex and 'scientific' than other optical instruments, they are of less interest to instrument historians, and perhaps regarded more in terms of fashion than technology. Spectacles are notably missing from studies of fashion or dress. Little is therefore known about the social and cultural context of spectacles and their normative relationship with attitudes to vision and the eye.

But spectacles were part of the larger continuum of attitudinal changes towards what was materially possible for the body. As other chapters in this book demonstrate, the arrival of new technologies coincided with changing beliefs about bodily alteration and the ability to 'correct' errant nature. As steel technologies became more prominent, they both accompanied and augmented the course of change. This chapter therefore makes three main claims. First, the design of spectacles changed. Second, either as a result of new technologies or of wider cultural shifts in understandings of the body, attitudes to eyewear also began to alter. Third, makers and consumers of eyewear faced new social and practical circumstances for which new products were required. New associations with learning and sagacity replaced connections of spectacle wearing with age and infirmity. As this happened, wearing spectacles moved from the intimate private sphere into the public arena.

Perhaps paradoxically, the focus here is upon the design and development of spectacle frames rather than lenses. Whilst, in a discussion relating to the growing importance of the eye, relegating the lens might seem strange, it can be argued that the greatest change both to eyewear and to attitudes towards the wearing of spectacles was in fact driven by the changing nature of frames. New metals, such as cast steel, revolutionised the manufacture of spectacle frames. Steel enabled the introduction

DOI: 10.1057/9781137467485.0009

of sprung arms, making them adhere to the wearer's head rather than balance on their nose. But as objects in themselves, steel-framed spectacles were aesthetically pleasing, and reflected the growing estimation of steel as an 'enlightened metal'. As the form of frames changed, both the ways in which spectacles were used and the meanings behind them began to shift. If not yet entirely fashionable, spectacles no longer needed to be hidden.

A cult of 'seeing'?

The second half of the seventeenth century brought changes to concepts of sight and of the eye as both a key sensory organ and a delicate scientific instrument. Optical instruments such as microscopes were the acme of enlightened enquiry and opened up a new world of micromechanisms for scientists to explore. New interests in empirical observation, especially in terms of observing 'animalcula' and the flow of blood, saw vast numbers of microscopes being sold and exported. Microscopes were, in Al Coppola's words, a 'transformative sensory prosthesis of the New Science'.[9] As the organic embodiment of this new spirit of enquiry, the eye fitted not only the ideals of a precision instrument but also the demands of polite society, for which taste and discernment were paramount. It was the fundamental organ through which men of taste explored, codified and appreciated their world. 'When we look on the eye as an Optical Instrument', opined the optician James Ayscough, 'it is all admiration!' The eyelids 'defended' the machine and 'served as a screen to shut out the light, while the soul is asleep'.[10] In his 1789 *Essay on Vision*, the noted instrument maker George Adams Junior proselytised about the 'evident manifestations of exquisite art and design [of the eye], every part elegantly framed and nicely adjusted and commodiously placed... so manifold are the blessings that we derive from this organ, that the human mind seems almost inadequate to the conception'. These were not cold, disinterested appraisals of ocular function, but enlightened celebrations of the eye as a wonderful machine, created and set in motion by the master maker unseen.[11]

Sight itself was no less important. Indeed, the eighteenth century was a golden age of sight, of literally 'seeing' the world anew. For individuals, sight was a blessing, with any diminution of clarity representing both a social and economic barrier. The instrument maker Benjamin Martin

DOI: 10.1057/9781137467485.0009

lamented the melancholy condition of those with poor sight, compromised in trade and commerce, useless in war and unable to maintain correspondence with friends.[12] More broadly, this was an age of both social and scientific observation, and of what Joanna Picciotto calls 'professional observers' – scientific and literary virtuosi who not only focussed upon new worlds through the lenses of their modish instruments but also used writing as a virtual instrument through which they could peer at their own social worlds.[13] This was the era of the 'gaze', with everything discernible and observable. Its effects were felt everywhere from science to medicine to punishment. The whole purpose of the Panopticon prison was, after all, to harness the power of the unseen or implied gaze to force prisoners into regulating their own behaviour.[14] Landscape architects like Lancelot "Capability" Brown transformed scenes of wild nature into a neat and harmonious 'succession of pictures' to encourage travellers to pause and reflect.[15] The eye could be led around a landscape, just as it could a painting, with 'eye-catcher' elements creating an impression upon the viewer.[16] Even verse and literature were profoundly visual, with romantic poets like Thomas Gray constructing mind pictures of idyllic pastoral scenes – his imagined evening landscape 'glimmering on the sight'.[17]

There was indeed a close semantic correlation between literal and notional sight, and eighteenth-century language was shot through with visual references and metaphors. Enlightened thinkers were encouraged, both literally and notionally, to open their eyes, broaden their outlook, vision or viewpoint. Satirists used sight or myopia to imply a want of understanding.[18] 'A Pair of Spectacles for Short-Sighted Politicians of 1789' ran the subtitle to one satirical pamphlet.[19] Other 'men of letters' used spectacles as a humorous metaphor to encourage greater insight or a change of viewpoint. In a letter to Jonathan Swift, Viscount Bolingbroke advised that the best path to follow in life was that of reason rather than custom. 'To be sure of doing this', he suggested, 'you must put on your philosophical spectacles'.[20]

As we have seen in other chapters, the eighteenth century saw a burgeoning market for instruments and devices for the body. This market was fed by a vibrant culture of metallurgical experimentation by small-scale artisans and specialist makers, using new materials to extend the tools and products of their trade.[21] Both as optical instruments in their own right and as specific corrective technologies of the body, spectacles were part of this process. The impact of these changes was felt in

DOI: 10.1057/9781137467485.0009

attitudes towards the causes of poor sight. Representations of spectacles in portraits, for example, reveal an ambiguity about their purpose. The connection between age and the diminution of sight was ubiquitous, but its significance less clear. On the one hand, spectacles were a metaphor for the learning and insight of the aged sitter, representing a life spent in study or contemplation. On the other, portraying an elderly sitter with spectacles either atop the head or balancing on the nose drew attention to the dereliction and decrepitude of age, suggesting the stereotypical myopia of the old fool – literally a lack of vision. Indeed, elderly people not requiring spectacles were regarded as exceptional. In March 1715 in Normandy, Monsieur Chesnar, aged 111, died, who 'was never ill and always read without spectacles'.[22] The death in 1740 of Mr Davis, a London centenarian, was remarked upon in the *Daily Post* since '*in particular* [my emphasis] his sight was so good that he could read the smallest print without spectacles'.[23] In this sense spectacles were almost a rite of passage into old age.

Age was certainly a defining factor in diagnosing and treating eye conditions, and assessing the need for spectacles. One user of the dubious 'Strengthening Eye Water' reported having lost nearly all sight but, since using the water for over fourteen years could now read without spectacles '*altho' he is now above 60 years of age*' [my emphasis].[24] The expectation was of the decline, diminution or even total loss of eyesight from the onset of middle age. Spectacles were often recommended for the young to preserve their sight in later life. Some opticians indeed used the patient's age as the primary factor in their prescriptions, and spectacles were sold by stages, or even decades of life. In April 1722, the optician John Marshall ('maker of optick glasses to his Majesty') 'set the age upon the frame' to prevent people using the wrong spectacles for their age.[25]

In relation to changing gender conceptions and ideals of masculinity, however, it is interesting to note an early eighteenth-century report of an apparent fashion for spectacles. In 1709, a correspondent to the *Tattler* noted an 'invented ambition' for fashionable short-sightedness amongst young society gentlemen, prompting them to continually peer at each other, and the opposite sex, through eyeglasses. 'At a lady's entrance into the playhouse, you might see Tubes immediately levell'd at her from every quarter of the pit and side boxes'.[26] Whilst this is a humorous note on the affectations of youth, the anomaly of young people with eyewear is plain. So strong were connections between age and ocular decline that its subversion was noteworthy.

DOI: 10.1057/9781137467485.0009

Nonetheless, by the 1720s, changes were taking place. For the king's personal optician, Edward Scarlett, the idea of spectacles based on age was outmoded and potentially injurious, encouraging people to use spectacles who did not need them, or to use examples entirely unsuitable for their eyes. Scarlett tailored his instruments to the individual, asking them to provide the focal distance, defined as 'the distance at which they [i.e. the person's regular spectacles] burn when exposed to the sun'.[27] This was a markedly different approach and one in line with shifts towards the importance of empirical observation. Some opticians positioned themselves as scientists engaged in continual experimentation to refine their products for individual consumers. Their shops were well equipped and located on busy shopping streets. James Ayscough was located in Ludgate Street near St Paul's, a bustling thoroughfare. The optician, spectacle and microscope maker John Cuff kept shop in Fleet Street, near the luxurious clothes shops and mercers patronised by the wealthy.[28] As the eyes became privileged, eyewear began to lose negative associations with the deleterious effects of age and become something that, if not yet overtly desirable, could be an acceptable and potentially modish accessory. The chapter now turns to spectacle design and development.

The design and marketing of spectacles

Early spectacles sat upon the bridge of the nose. They consisted of two round lenses, usually mounted in leather frames and joined together with a fixed bridge piece. They had no arms, and the nosepiece was not generally sprung, although they were light, portable and easy to put on and take off quickly. The design of these 'nose spectacles' was simple, easy to produce and replicate, and generally cheap to purchase. The first developmental change in the form of spectacles' design came in the sixteenth century with the invention of 'slit' or 'split' nose spectacles in either horn or tortoiseshell frames. These were so called because of the stiff bridge or bow piece across the nose, which contained up to four decorative slits, depending on the price of the items. Lenses were ground using iron tools and employing methods developed by Venetians for mirror glass, and allowed variation for long- and short-sightedness.[29]

In the late seventeenth century, changes were made both to lens and frame manufacture. The introduction of leather-framed spectacles rendered them cheaper and more durable. New methods of lens

DOI: 10.1057/9781137467485.0009

grinding and surfacing on brass tools by London artisans improved the utility of spectacles alongside other types of eyewear.[30] These spectacles, often referred to as the 'common sort', were ubiquitous across Britain and across social classes by the seventeenth century. Mercers and general traders often sold them, along with travelling 'specialists', and they ranged in price from a few pence to several shillings. The 1694 account books of Sir Harbottle Grimston, the wealthy London lawyer and politician, for example, contain an entry for 'two pairs of leather spectacles' costing five shillings in total.[31] In one seventeenth-century Welsh mercer's shop, however, the same price would procure no fewer than fourteen pairs of spectacles with cases – significantly cheaper than Grimston's purchase and perhaps reflecting cheaper materials or lenses, as well the provincial location of the mercer.[32] Nose spectacles were specifically for long-sightedness, that is, for use at close range, and were utilitarian items designed simply for reading. Although spectacles and lenses were available to enable the short sighted to see at a greater distance, they were not intended to be worn continuously. As such, the wearing of spectacles was inherently private; to be seen wearing them in public would be exceptional. In some parts of Europe, spectacles could become fashionable though. A French visitor to seventeenth-century Madrid was surprised by the apparent vogue for spectacle wearing amongst young women, balanced on the nose but attached to the head. She was told that this was 'done to make them look serious', and that some ladies were only without their spectacles when abed.[33] There is no evidence to suggest that this trend was widespread.

Arguably the most important developments in design accompanied the growing introduction of metals into manufacture. The influence of iron and steel in fact predated the introduction of cast steel. One of its first effects on spectacle making was the transformation of the nose bridge from a rigid to a sprung piece.[34] More comfortable for the wearer, spectacles could now adhere more firmly to the wearer's nose. Perhaps the biggest single change occurred in 1727, however, when the London spectacle maker Edward Scarlett supposedly created the first spectacles with arms.[35] Using rigid steel pieces with padded hoops at either end, Scarlett's spectacles obviated the continual need to hold them in place with one hand. These were later developed to include hinges and loops to fit around the ears, and were available in various frame materials.[36]

Other metals altered the form and appearance of spectacles. Frames of silver, gold, iron, bronze and nickel-silver, a type of silver alloy, appeared

DOI: 10.1057/9781137467485.0009

FIGURE 5   *Scarlett-type spectacles, Edward Scarlett, c. 1730*
Source: © The College of Optometrists, British Optical Association Museum.

in advertisements.[37] Silver spectacles had been used by Elizabethan elites, while cheap wire-framed examples were commonly used lower down the social scale. Nonetheless, the continual environment of metallurgical innovation in the eighteenth century brought spectacle making into a continuum of modification and experiment. In fact, as frames became more ornamental, attitudes towards both the wearing of spectacles and their appearance shifted. If it was not yet fashionable to wear spectacles, then it was at least unfashionable to wear ugly ones. If they were necessary, then, they should at least be elegant.

From mid century, more elaborate spectacle frames emerged, especially in bespoke items for the well heeled. The spectacle frames of wealthy elites could be made of precious metals such as gold and silver, with frames set with precious stones.[38] The new vogue for 'cast steel' jewellery in the second half the eighteenth century certainly impacted upon spectacles design, and frames made of this lustrous material even had potential as fashion items. Rosenthal, for example, points to the presence of armorials and other status indicators on spectacle frame, reflecting their potential to be show items – something to be displayed rather than hidden.[39]

In fact, advertising laid increasing emphasis upon spectacles frames. Earlier advertisements were likely to privilege the technology of the lenses and methods of grinding.[40] From around 1730, though, the

DOI: 10.1057/9781137467485.0009

aesthetics of the frame gained importance. This, it could be argued, drew attention to spectacles rather than encouraging concealment. In 1756, the London instrument maker Benjamin Martin introduced his 'Martin's Margins' spectacles. Complete with sprung 'sides' (the technical name for arms), these had large circular frames around the eyes with a thick inner central band of ivory. In Martin's own words, they had 'partially obstructed apertures so that the eyes were not overloaded with light' and lenses that 'tilted inwards so that the axes of the eye converged on the object of regard'.[41] In part, they sought to limit the amount of light reaching the eyes, believed to be debilitating to sight, and also acted to force the eyes into focus. But their radical design also drew attention both to the spectacles and to the wearer's eyes. Despite their appearance, they were not unpopular.

If spectacles were not necessarily fashionable, then, it clearly *mattered* what they looked like. The optician Joseph Linnell of Ludgate Street in London sold spectacles and reading glasses set 'in neat and commodious frames'.[42] Samuel Whitford, again of Ludgate Street, made and sold spectacles 'in neat and light frames' and 'of the newest and best construction'.[43] One of the best examples of the range and style of spectacles during this period is that of the 1765 catalogue of instrument maker Henry Pyefinch of Cornhill, London.[44] At the very top end of Pyefinch's stock were 'brazil pebble' spectacles in a double-jointed frame, for the substantial sum of one pound and eighteen shillings. Metallic frames were generally amongst the most expensive, and frame form also determined price. Whereas a pair of 'Spectacles for the nose' could be procured for a shilling, several pairs of steel and silver spectacles in double-jointed frames cost more than a pound.[45] In the mid range were items like 'best silver spectacles for the temples' costing fifteen shillings, with steel examples generally priced slightly lower.[46] Other materials included tortoiseshell and horn. More elaborate examples had frames of mixed materials, such as concave pearl and silver spectacles costing twelve shillings. Such options highlight the importance of appearance as well as functionality. Lenses were not simply mounted in workaday frames; rather, the consumer had choices to suit their pocket and also their preference.

The importance of appearance is reinforced in the changing nature of spectacle cases. In advertisements, the materials and appearance of cases was a selling point. Although the wealthy presumably always had cases to befit their station, early cases tended to be functional rather than decorative, made from basic materials such as bone, wood or leather to protect

DOI: 10.1057/9781137467485.0009

the items from damage.[47] By the 1740s, however, cases were becoming ever more elaborate. A pair of silver temple spectacles in the Zeiss collection dated to 1740 was housed in a matching silver and filigree case.[48] Again, elaborate spectacles and cases were targets for thieves. In July 1741, for example, several defendants were indicted for stealing items including a pair of spectacles with a shagreen case.[49] In 1757, Thomas Dumble was indicted for stealing six pairs of temple spectacles 'and fishskin cases', as well as various frames, from the London optician Leonard Ballit.[50] Although available since the seventeenth century, shagreen, made from sharkskin, was a material of the moment, utilised for many purposes including lining for cases and boxes.[51] In one sense cases could simply be regarded as trophy items and as an elaborate and fashionable means of disguising the fact that the owner wore spectacles. In this sense the case, rather than the contents, was the fashionable item. But a showy shagreen or polished steel case drew attention to, and emphasised, its contents, in effect celebrating the spectacles as fashionable and modish accoutrements.

One relation of spectacles that was more firmly associated with fashion was the quizzing glass or 'lorgnette'. This consisted of either a pair of spectacle with a handle or stem, or a single lens worn on a ribbon and held up when the user wished to scrutinise someone or something. Again, these items could be extremely ornate, with high-end examples enamelled and gilded, or set with precious stones.[52] These were show items, designed to be useful but also highly visible. By 1800, they were all the rage, especially with young men eager to survey the opposite sex. As a letter to the *Female Spy* in 1800 suggested, fashionable nearsightedness had made a comeback, with the writer describing how the 'black ribbon of my pendant quizzing glass contrasts the white of my high cravat'.[53] Satirists like Matthew Darly lampooned the affectation and narrow vision of young men who could only squint at the world through the lenses of their lorgnettes.[54]

As spectacles were absorbed into broader themes of polite bodily display, the trade and retail of spectacle making also began to alter. This is reflected in the adoption of the language of politeness or fashionability by advertisers, but also by the recasting of optical instrument makers as 'opticians', emphasising their role as natural philosophers, as well as optical specialists. The term 'optician' was first coined in 1672 (allegedly by Sir Isaac Newton), referring to a general specialist in the science of optometry. But it was arguably not until the mid eighteenth century that 'optician' came to be applied to, and used by, makers and retailers of spectacles.

DOI: 10.1057/9781137467485.0009

Opticians' trade cards, in line with a general shift towards polite style and language in advertising, included fashionable motifs such as rococo surrounds. Thomas Ribright, optician to the Prince of Wales, depicted a pair of spectacles below an elaborate rendering of the royal coat of arms, a heavy rococo surround and various optical instruments in the process of being used by putti.[55] The scientific instrument maker John Gilbert similarly framed his instruments with rococo, but also included an image of a 'scientist' gazing into a sextant in some imaginary exotic location, his ship anchored in the background.[56] Others, as did razors makers, embellished their advertisements in the new rhetoric of science. The maker of 'Smith's Patent Dilating Spectacles' was confident of their superiority in 'philosophical principle as well as their mechanical formation'.[57] John Yarwell's 'True Spectacles' were constructed in a 'manner approved by the Royal Society'.[58] In emphasising the scientific credentials of their wares, makers sought to encourage a specific, technologically savvy and genteel customer to a prosaic product. It is worth noting, though, that as small, quotidian items, spectacles also fitted what might be considered 'impulse buy' items (like playing cards, lottery tickets and even compound medicines) and were found stocked by diverse retailers of 'toys'.

Whether more people actually wore spectacles is hard to establish, since there is little quantitative evidence for spectacle use. Whilst advertising became more prominent, the inclusion of spectacles amongst the stock lists of instrument makers makes it difficult to assess how far spectacles were bound up with the broader vogue for instruments. There is some evidence that the use of spectacles as a sight preserver was increasing – a trend that alarmed George Adams. In the preface to his treatise on vision he sought to do away with this 'general prejudice in favour of spectacles' that caused 'numbers to use glasses before they could be of any essential service'.[59] The Hereford sellers of Bradberry's Patent Spectacles in 1795 claimed to have '[l]ists of some thousands who have had them in the course of the last twelve months'.[60] Given the range, price and availability of eyewear, it seems plausible that they were utilised by large numbers of the population and across society.

Function

Having identified changes in spectacle design during the eighteenth century, and the privileged position of the eye as a key organ of sense,

DOI: 10.1057/9781137467485.0009

the remaining question is the relationship between spectacle wearing and the bodily self-fashioning central to this book. In other chapters, the connection between fashion and technology is easy to find. Steel jewellery transformed the market, demonstrating how an innovative new industrial material became the acme of taste and fashion.[61] Cast steel razors likewise helped gentlemen convey elegance and refinement. But spectacles are harder to fit into this framework. Except for short periods, such as the earlier example of their adoption by young beaus as an affectation, there was by no means a vogue for spectacle wearing in the same way as, for example, steel jewellery. They were worn by people who needed them, rather than those who simply desired a certain 'look'. Nevertheless, spectacles gradually changed from functional items to acceptable adornments. For the first time they became, literally, 'eyewear'.[62] As this occurred, spectacle wearing moved from the private to the public realm as new circumstances and social contexts demanded either the adaption of existing designs or the creation of new ones.

One of the main changes surrounded the figurative linkage of spectacles with learning, and especially reading. The early modern period saw an unprecedented amount of printed material, from pamphlets and newspapers to books. To be sure, reading could be a private and solitary affair. On the one hand, the libraries of large houses encouraged reading as an immersive experience, the individual closeted away with their books. On the other, cheap newspapers and coffee-house culture brought an increasingly public dimension to reading. Historians have long acknowledged the importance of the coffee house as a site for the consumption of news and literature.[63] Most commonly, coffee houses are viewed as a repository of newspapers and periodicals, but they often contained surprisingly well-stocked libraries of other types of literature. Markman Ellis argues that journals, works of literature and fiction, poetry and works of reference all featured in coffee-house libraries in large towns across England, and not just in the capital.[64] Reading in this context was public, in the sense that it was undertaken in a public environment, but also often a shared experience, with reading aloud considered a social skill.[65] Spectacles were a part of this culture as an expedient to reading. But they also embodied the spirit of learning.

Unsurprisingly, the new vogue for reading actually contributed to the need for spectacles. Reading densely printed works in low light led to painful eyestrain – as Samuel Pepys had found to his cost in the 1660s. Those engaged in close work regularly suffered eye complaints. In his

DOI: 10.1057/9781137467485.0009

1790 treatise on ocular conditions, William Rowley commented upon the injurious effects upon the eyes of certain trades. Watchmakers, who 'work on minute objects', often needed parallel glasses, adopted at an early age to preserve their sight.[66] Young ladies who undertook fine needlework, he argued, risked painful and watery eyes or contracted pupils.[67] Not without a dash of humour, he also suggested that 'too close an attention to our modern amusement cards' should be avoided.[68]

Depictions of spectacles in portraiture appear to support changes in attitudes towards eyewear – at least among elites. By the later eighteenth century, sitters were more likely to be depicted with their spectacles. A 1764 portrait by Mary Black, of the physician Messenger Monsey (d. 1788) depicts the subject in repose, a book in his lap and a pair of temple spectacles in his hand. Another 1784 portrait by Robert Fulton

FIGURE 6   *R. Fulton, Portrait of Robert Harris, 1784*

Source: © The College of Optometrists, British Optical Association Museum.

DOI: 10.1057/9781137467485.0009

of the untraced sitter Robert Harris, reproduced in Figure 6, similarly shows a seated figure holding a book and a pair of modish steel-framed spectacles.

A portrait assumed to be Benjamin Franklin depicts the then-US senator closely scrutinising a newspaper through a pair of temple spectacles. On one level, depicting spectacles together with books emphasised their utility in reading. Symbolically, however, they also acted as a metaphor for the sitter's bookishness and scholarship. In the portraits of Monsey and Harris, the viewer is presented with figures gazing away, apparently brooding on the book they have just read. Both spectacles and book are signposts to the personality and habit of the sitter. Franklin, by contrast, is depicted in the act of reading, his spectacles prominent. Attention is drawn to his eyes and to his gaze, his posture and expression emphasising the intensity of his scrutiny. The fact that he is reading a newspaper may imply his grasp of current events. In each case the positioning of eyewear is prominent; no attempt is made to disguise their use. The fact

FIGURE 7   *Stephen Elmer, 'The Politician: Benjamin Franklin', c. 1780*
Source: © The College of Optometrists, British Optical Association Museum.

DOI: 10.1057/9781137467485.0009

that each sitter was clearly fairly affluent provides further evidence that spectacles were becoming acceptable accessories of the savant.

As in the literary caricatures noted above, spectacles found use as visual shorthand in political satires, in which they could imply deficiency. The politician Edmund Burke was often caricatured wearing spectacles. As Nicholas Robinson points out, they rendered Burke easily recognisable in a group caricature (itself suggesting the distinctiveness of the items), but also implied defective sight – again, a visual pun on the supposed lack of political vision. Burke's spectacles could symbolise his point of view. He might be pictured as having removed them, not wishing to see that which was before him. Alternatively, others might look through his spectacles to see the world as he did.[69] Here again, the depiction of spectacles was bound up in a number of literal and symbolic meanings. Whether the appearance of spectacles in visual art represents anything like the valorisation of eyewear, or instead simply reflects a turn towards informality in portraiture, is open to question. Whatever the reason, their apparently increasing appearance in pictures was almost certainly deliberate.

There is certainly evidence too that spectacle designs altered to accommodate changing social environments. Perhaps most obvious was the development of 'wig-spectacles'. By the 1770s, wigs were the height of enlightened fashion for both sexes. As Margaret Powell and Joseph Roach have argued, the preparation, styling and appearance of hair was part of the 'social performance of everyday life'.[70] Hair, especially in the form of large and elaborate wigs, was itself both a performance and a public statement with its own rituals and meanings. This was, after all, the age of the society ball, the theatre and of public promenading around the newly built shopping streets of fashionable towns such as London and Bath. As the sites of social gathering proliferated, so the need for a public 'face' was paramount. Wigs were, for their wearers, a virtual dialogue of politeness and sociability. Given the emphasis upon harmonious self-presentation, it might seem surprising that items as prosaic as spectacles might have a function. Nonetheless, makers such as James Ayscough began to experiment with double-jointed and extended sides, allowing spectacles to fit around the wig. Wig spectacles were functional, to be sure. They addressed a problem for wig wearers insofar as regular temple spectacles were unsuitable, while nose spectacles were inconvenient. But they were clearly aimed at, and worn by, a narrow section of society. These were highly visible items. Rather than being tucked inside the wig, their long arms wrapped incongruously around its circumference. In

DOI: 10.1057/9781137467485.0009

private, these types of spectacles would be redundant; a wig could simply be removed. In public, they were a self-conscious statement, a highly visible articulation of learning and sagacity. In being absorbed into this new vogue for hair, and the broader importance of the head as a site of display, the social meanings of spectacles were clearly shifting.

In other cases, changes to eyewear were more pragmatic and resulted from either changing views of ocular health or new circumstances to which the eye was becoming exposed. The idea of eyeglasses as a form of protection, including tinted lenses, can be traced to the eighteenth century. Extremes of light or dark were believed to injure the eye. George Adams was clear in his view that sitting in either a gloomy room or a blaze of light was best avoided, as was going from one room to another too rapidly. Even strong colours could be detrimental to sight.[71] Spectacles with lenses either darkened or coloured were commonplace. In 1797, a patent was granted for a new type of spectacles with side visors – known as 'D spectacles'. They were designed to limit the amount of light entering the eye from the front and sides. By the early 1820s, an entirely new sensory experience – the railway – also stimulated changes to protective eyewear. Sitting in open carriages, the passengers precariously perched behind early locomotives were assailed by a barrage of steam, dust and sparks as they barrelled along at breakneck speeds of up to twelve miles per hour! Here, the protective function of 'D' spectacles was developed and reinforced. Becoming known as 'railway spectacles', they offered some protection against the elements, although their glass lenses were liable to shatter, causing potentially more damage to the eye than the railway journey itself.

Whether spectacle wearing was ever mimetic is unclear. A type of spectacles with green-tinted lenses was apparently associated with a Venetian actor Carlo Goldoni, although there is no evidence for deliberate endorsement or indeed for the extent of their popularity in England.[72] Except for fleeting reports, it is unlikely that people who did not need them generally wore them either as an affectation or to imitate others. Instead they were part of a range of corporeal devices that could be deployed to fashion the body according to given social circumstances.

Conclusion

Spectacles were an important technology of the body in the eighteenth century. As optical instruments, they were part of a scientific, medical

DOI: 10.1057/9781137467485.0009

and intellectual milieu that called for continuous experiment and development to expand the material boundaries of production. Spectacle frame design and function altered dramatically over the course of the eighteenth century. The springiness of newly discovered metals such as cast steel made frames lighter, thinner and more comfortable, while metal frames in general were more durable than their horn or leather predecessors. Spectacles with arms removed the need for wearers to continuously hold them in place, freeing their hands and potentially making reading more comfortable. The relationship between spectacles and metallurgical technologies, then, was a close one.

But spectacles were enmeshed within broader cultural changes both to eyesight and attitudes to eyewear. Whereas in previous centuries spectacles connoted deficiency – almost disability – they now became tolerable, if not exactly fashionable. The act of wearing spectacles changed from private to public, partly due to the rising social function of reading, both for polite edification and as a marker of sensibility. The popularity of literacy, driven by coffee house and newspaper culture, and cheap print, transformed reading from a private to a public activity. As such, spectacles were functional but could also symbolise the sagacity of the wearer. To wear them was not only to participate in the physical act of reading but also to engage in the wit, sociability and micropolitics of the coffee house. This public role of spectacles carried over into the specific development of items such as 'wig spectacles', designed to allow myopic savants to attend society balls and functions in comfort and safety.

As attitudes to eyewear changed, spectacles embodied the new vogue for sight. In portraits, spectacles drew attention to the wearer's eyes and, implicitly, their gaze or insight. Spectacles were optical instruments, but they were also facilitators for the acquisition and dissemination of knowledge. Through them, men of letters engaged with the latest scientific discourses and intellectual debates. The decision to be depicted in spectacles was therefore self-conscious and loaded with meaning. In this sense they were part of a dialogue of polite self-fashioning and presentment that acted as a lingua franca through which other like-minded individuals could identify them.

Whilst the question of fashion is debatable, spectacles were undoubtedly part of a changing technological and cultural landscape, one in which the body was a site of experiment and innovation. As new technologies altered the form of assistive devices of various sorts, so attitudes to the conditions previously considered as socially undesirable or

DOI: 10.1057/9781137467485.0009

limiting could also alter. As elsewhere, it is difficult gauge the extent to which new metallurgical technologies were primary drivers of change, or whether they facilitated developments already in train. In either case, spectacles were part of an ever-shifting relationship between the body, technology and culture in the eighteenth century.

Notes

1 Advertisement, 'Superfine Crown Glass', *Whitehall Evening Post or London Intelligencer,* 27 March 1750.

2 Ibid.

3 Richard Corson, *Fashions in Eyeglasses from the Fourteenth Century to the Present Day* (London: Peter Owen, 1967); J. William Rosenthal, *Spectacles and Other Vision Aids: A history and guide to collecting* (London: Norman Publishing, 1996); Wolf Winkler, *A Spectacle of Spectacles: Exhibition Catalogue* (Edition Leipzig, 1988).

4 Hugh Barty-King, *Eyes Right: The Story of Dollond and Aitchison, 1750–1985* (London: Quiller, 1986).

5 Moritz Von Rohr, 'Contributions to the History of the Spectacle Trade from the Earliest Times to Thomas Young's Appearance', *Transactions of the Optical Society,* 25:2 (1923–4), 41–72; Thomas H. Court and Moritz Von Rohr, 'On the Development of Spectacles in London from the End of the Eighteenth Century', *Transactions of the Optical Society,* 30:1 (1928–9), 1–12.

6 Ralf Willach, *The Long Route for the Invention of the Telescope* (Philadelphia: American Philosophical Society, 2008).

7 For examples see A.D. Morrison-Low, *Making Scientific Instruments in the Industrial Revolution* (Aldershot: Ashgate, 2007); Larry Stewart, *The Rise of Public Science: Rhetoric, Technology and Natural Philosophy in Newtonian Britain, 1660–1750* (Cambridge: Cambridge University Press, 1992); Jan Golinski, *Science as a Public Culture: Chemistry and Enlightenment in Britain, 1760–1820* (Cambridge: Cambridge University Press, 1992).

8 John Millburn, *Adams of Fleet Street: Instrument Makers to King George III* (Aldershot: Ashgate, 2000); Anita McConnell, *Jesse Ramsden (1735–1800): London's Leading Scientific Instrument Maker* (Aldershot: Ashgate, 2007); Gloria Clifton and Gerard L'Estrange Turner, *Directory of British Scientific Instrument Makers, 1550–1851* (London: Zwemmer/National Maritime Museum, 1995); James A. Bennet, 'Shopping for Instruments in London and Paris' in Pamela H. Smith and Paula Findlen (eds), *Merchants and Marvels: Commerce, Science and Art in Early Modern Europe* (London: Routledge, 2002); Lissa L. Roberts, Simon Schaffer and Peter Dear (eds), *The Mindful Hand: Inquiry and Invention from the Late Renaissance to Early Industrialisation* (Chicago: University of Chicago Press, 2008).

DOI: 10.1057/9781137467485.0009

9   Al Coppola, "'Without the Help of Glasses": The Anthropocentric Spectacle of Nehemiah Grew's Botany', *The Eighteenth Century*, 54:2 (2013), 264.

10   James Ayscough, *A Short Account of the Eye and Nature of Vision* ... (London: Printed by E. Say, 1754), 3.

11   See David C. Lindberg and Ronald L. Numbers (eds), *God and Nature: Historical Essays on the Encounter Between Christianity and Science* (Berkeley: University of California Press, 1986); Jon Hedley Brooke, *Science and Religion: Some Historical Perspectives* (Cambridge: CUP, 1991).

12   Benjamin Martin, *An Essay on Visual Glasses (commonly called Spectacles)* ... (London: Printed for the Author 1758), 4.

13   Joanne Picciotto, 'Optical Instruments and the Eighteenth-Century Observer', *Studies in Eighteenth-Century Culture*, 29 (2000), 124.

14   See Janet Semple, *Bentham's Prison: A Study of the Panopticon Penitentiary* (Oxford: Clarendon Press, 1993).

15   See various essays in Denis Cosgrove and Stephen Daniels (eds), *The Iconography of Landscape: Essays on the Symbolic Representation, Design and Use of Past Environments* (Cambridge: CUP, 1988); also John Phibbs, 'The View-Point', *Garden History*, 36:2 (2008), 215–7.

16   Ibid., 217.

17   Thomas Gray, *An Elegy Wrote in a Country Churchyard* (London: Printed for R. Baldwin, 1751).

18   For studies of visual metaphor in the eighteenth century, see Barbara Stafford, *Imaging the Unseen in Enlightenment Art and Medicine* (Cambridge, Ma: MIT press, 1991); Martin Kemp, *Visualisations: The Nature Book of Art and Science* (Oxford: OUP, 2000).

19   See Anon, *A Pair of Spectacles for Short-sighted Politicians* (London: Printed for J. Williams, 1765); Anon, *Alfred Unmasked or the New Cataline; Intended as a Pair of Spectacles for the Short-sighted Politicians of 1789* (London: Printed for R. Faulden, 1789); Anon, *Now or Never or A Looking Glass for the Representatives of the People and a Pair of Spectacles for the Freeholders of England* (London: 1702).

20   British Library, Add. MS 4805, Letter Henry St. John, 1st Viscount Bolingbroke to Jonathan Swift, 27 September 1729.

21   Chris Evans and Alun Withey, 'An Enlightenment in Steel? Innovation in the Steel Trades of Eighteenth-Century Britain', *Technology and Culture*, 53:3 (2012), 550–1.

22   Anon, 'From the Paris Gazette', *Weekly Journal with Fresh Advices Foreign and Domestick*, 26 March 1715.

23   'London', *Daily Post*, 14 April 1740.

24   Advertisement, 'The Strengthening Eye Water', *Post Man and the Historical Account*, 19 May 1719.

25   Advertisement, 'A Magick Night Watch to Shew the Hours of the Night", *Daily Post*, 20 February 1722.

DOI: 10.1057/9781137467485.0009

26 Anon, *Tattler,* Tuesday, 4–6 October.

27 Ibid.

28 Bodleian Library, John Johnson Collection, MS Scientific Instruments, 1, 'John Cuff, Optician, Spectacle and Microscope Maker'.

29 Court and Von Rohr, 'On the Development', 7–8.

30 Ibid., 8.

31 Andrew Browning (ed.), *Extracts from Account Books of Sir Harbottle and Sir Samuel Grimston, 1683–1700, part 5, Document 2,* 175, http://o-www. englishhistoricaldocuments.com.lib.exeter.ac.uk/document/view. html?id=3470.

32 National Library of Wales MS SD/1693/208, Probate inventory of William Lloyd of Carmarthen, January 1693.

33 C.S. Flick, 'Spectacles as Badges of Rank and Learning', *Vision,* 3:1 (1949), 36.

34 J. William Rosenthal, *Spectacles and Other Vision Aids: A History and Guide to Collecting* (San Francisco: Normal, 1996), 41.

35 Some debate surrounds whether Scarlett actually invented or patented the items, or whether he was simply the first to advertise them.

36 Rosenthal, *Spectacles,* 41.

37 N.J. Kook, 'Metal Frame Materials', *MOI Symposium Paper,* March 1978, 17–18.

38 Ibid.

39 Rosenthal, *Spectacles,* 41.

40 For examples, see Advertisement, 'G. Willdey and T. Brandreth at the Archimedes and Globe', *Daily Courant,* 24 March 1707; Advertisement, 'Spectacles Improved to Perfection by J. Marshall', *Daily Courant,* 8 May 1707.

41 Quoted in Rosenthal, *Spectacles,* 106.

42 Collection of the British Optical Association Museum, London, Advertisement, Joseph Linnell, 'The Original Shop for Crown Glass Spectacles", undated, c. 1774.

43 Collection of the British Optical Association Museum, London, Advertisement, 'Samuel Whitford, Optical, Mathematical and Philosophical Instrument Maker', undated, c. 1775.

44 Collection of the British Optical Association Museum, London, Anon, *A Catalogue of Optical, Philosophical and Mathematical Instruments Made and Sold by Henry Pyefinch at the Golden Quadrant, Sun and Spectacles* (London: 1765?).

45 Ibid.

46 Ibid.

47 For examples, see Winkler, *Spectacle of Spectacles,* 59, 66.

48 Ibid., 71.

49 OBPO, 1July 1741, trial of Elizabeth Eccles Richard Eades Elizabeth Jones alias Carnaby Mary Eccles alias Pugh, available at http://www.oldbaileyonline.org/ browse.jsp?id=t17410701-11-defend111&div=t17410701-11#highlight, accessed 13 December 2014.

DOI: 10.1057/9781137467485.0009

50 OBPO, 14 September 1757, trial of Thomas Dumble, available at http://www. oldbaileyonline.org/browse.jsp?id=t17570914-21-defend238&div=t17570914-21#highlight accessed 13 December 2014.

51 See Stuart Talbot, 'Shagreen and Fish Skin: On Scientific Instruments, Their Cases and Etui, circa 1700–1800', *Bulletin of the Scientific Instrument Society,* 123 (2014), 10–27.

52 See the examples in Winkler, *Spectacle of Spectacles,* 104–5, 109–10.

53 'The Female Spy', *Oracle and Daily Advertiser,* 31 December 1800.

54 Matthew Darly, *The Optical Contrast* (London: 1771).

55 Science Museum, MS 1951-785/63, Trade card of Thomas Ribright, c. 1753–72.

56 Science Museum MS 1934-0096, Trade card of John Gilbert, undated, early nineteenth century.

57 Advertisement, 'Smith's Dilating Patent Spectacles', *Morning Herald and Daily Advertiser,* 29 September 1784.

58 Collection of the British Optical Association Museum, London, Advertisement, 'Yarwell's True Spectacles', undated, late eighteenth century.

59 Martin, *Essay on Vision,* 1.

60 Anon, *By Permission of the Worshipful the Mayor, Now Exhibiting at Miss Newton's, Milliner, The Powers of Imagination or the Senses Deceived* (Hereford: Printed by D. Walker, 1795).

61 Evans, 'Crucible Steel', 83–4; See also Marcia Pointon, *Brilliant Effects: A Cultural History of Gem Stones and Jewellery* (New Haven, CT: Yale University Press, 2010).

62 The term is used figuratively here; it was not contemporary and did not appear until the twentieth century.

63 For examples, see Helen Berry, *Gender, Society and Print Culture in Late Stuart England: The Cultural World of the 'Athenian Mercury'* (Aldershot: Ashgate: 2003); Steven Pincus, '"Coffee Politicians Does Create": Coffee-Houses and Restoration Political Culture', *Journal of Modern History,* 67 (1995), 807–34.

64 Markman Ellis, 'Coffee-House Libraries in Mid-Eighteenth-Century London', *The Library,* 10:1 (2009), 12, 27–30.

65 Ibid., 32.

66 William Rowley, *A Treatise on One Hundred and Eighteen Principle Diseases of the Eyes and Eyelids* (London: 1790), 355.

67 Ibid.

68 Ibid., 357.

69 Nicholas K. Robinson, *Edmund Burke: A Life in Caricature* (New Haven, CT: Yale University Press, 1996), 152, 161.

70 Margaret K. Powell and Joseph R. Roach, 'Big Hair', *Eighteenth-Century Studies,* 38:1 (2004), 81.

71 Adams, *Essay on Vision,* 97.

72 A. Von Pflugk, *Farbige, insbesondere grune Glaser als Augenschutz,* (n.p.: Publisher unknown: 1929), Plate II, no. 6.

DOI: 10.1057/9781137467485.0009

Coda: Surgical Instruments and Bodily Transformation

Abstract: *It is easy to overlook the role surgical instruments in transforming the body. Withey details the impact of cast steel both upon the design of eighteenth-century surgical instruments and upon operating techniques. Whilst such instruments were not used by people upon themselves, the properties of cast steel allowed developments in instrument design, which, in turn, were marketed on the basis of reducing pain for surgical patients. Withey uncovers the markedly similar themes of technological innovation, marketing and advertising, which link surgical instruments to other technologies discussed in the book. Like razors, for example, surgical instruments were made by specialists, who emphasised their metallurgical and practical knowledge. This chapter adopts a novel approach to the relationship between bodies and technology, by exploring bodily transformation by others, rather than individuals.*

Withey, Alun. *Technology, Self-Fashioning and Politeness in Eighteenth-Century Britain: Refined Bodies.* Basingstoke: Palgrave Macmillan, 2016. DOI: 10.1057/9781137467485.0010.

This book has so far focussed upon the relationship between technological innovation and self-transformation. Razors, spectacles, instruments of personal grooming and postural devices could all be bought and used by individuals without recourse to a practitioner. Each device offered people the opportunity to shape their own bodies, whether to 'correct' disability or deformity, or to 'improve' posture, appearance and so on. But there was another group of instruments that were materially affected by new metallurgical technology, and which afforded new possibilities for the bodily alteration. In the second half of the eighteenth century, surgical instruments were transformed by cast steel. As was the case with razors, the opportunities afforded by the physical properties of steel brought changes in the design of many surgeons' instruments, in turn affecting operative techniques. This type of instrument clearly follows a different trajectory to the others discussed above. Amputation knives, saws, scalpels and other similar instruments were not designed or marketed for the public. They were not used by individuals upon themselves and, indeed, people went to great lengths to avoid them. Neither can they be considered in terms of the construction of a 'polite' body. In that sense they perhaps appear incongruous here.

But surgical instruments, and their makers, were still important vectors for the transformation of the body. Whilst caution must be taken to avoid a teleological tale of progress and improvement, sharper and more durable instruments shaped surgical practice. The ways in which makers marketed their products also bears comparison with other technologies in this book. There were also similarities and indeed crossovers between surgeon's instrument makers and other trades. Both makers and practitioners were engaged in continual experimentation with and refinement of their products. This chapter, therefore, offers a brief discussion of the effects of new technologies upon surgical technique and bodywork, and also the place of surgeons' instrument makers within the broader technological milieu of the late eighteenth century.

The literature on eighteenth-century medicine and medical education is now huge, but far less attention has been paid to the instruments wielded by surgeons, and their use and impact upon the patient experience.[1] Since Ghislaine Lawrence's 1992 essay bemoaning the dearth of surgical instrument historiography, the topic has attracted the attention of historians, however.[2] John Kirkup has written extensively about the history of surgical instrument manufacture, highlighting the changes wrought by steel in the eighteenth century.[3] Elizabeth Bennion has

DOI: 10.1057/9781137467485.0010

likewise illustrated the development of instrument design, while others, such as Christopher Booth, chart the effects of metallic substances in new healing devices such as 'metallic tractors'.[4] More broadly, studies of instrument making have highlighted the often-complex relationship between material, manufacture, design and application. The relationship between surgical instrument design and use, however, is less clear. Recent studies of dissection and anatomy are firmly focussed upon both theory and practice, but not the material culture of instruments and their place in the procedure.[5] In particular, the effect of new technologies upon the *outcomes* of surgery remains obscure. Tina Kausmally's essay on archaeological evidence for dissection in early modern human remains suggests possible technique, but does not speculate on the types of instruments used. Establishing whether the knives used were curved, straight, serrated and so on would be revealing about how instrument design shaped operative techniques. Likewise, the recent historiography of instrument making actually says little about surgical instruments, although it could be argued that they were first and last 'scientific instruments'. Although viewed partly as belonging to a 'mechanical' part of science, surgical instruments were essential to the empirical discovery and classification of the body machine.[6] Whilst it would be difficult to argue that steel instruments reduced patients' perceptions of the agony and discomfort of surgical procedures, or influenced people's desire to fashion their own bodies, they remain an overlooked element in both the conceptualising and transforming of the eighteenth-century body.

The market for instruments

The surgeon's kit held an ambiguous position within medicine. For patients, the merest sight of a knife was apt to make them quake. Surgeons like Peter Clare noted that many would rather seek the dubious services of a quack than subject themselves to the agony of the blade.[7] At the same time, the practice of surgery (as opposed to theory) was medicine's lesser art. Over time and with practice, it was argued, virtually anyone with basic anatomical knowledge and a dextrous hand could wield a knife. Without the requisite theoretical understanding of the body, however, this was a mere 'mechanical part of the profession'.[8] In this sense, amputation knives, scalpels and saws were little more tradesmen's tools or, as

DOI: 10.1057/9781137467485.0010

the medical author Johannes Van Horn put it, 'inanimate servants and necessary companions in the business to be effected'.[9]

Driving demand for good quality instruments was the growth of medical training. Across Europe, there was a marked increase in the numbers of students studying surgery. In the third quarter of the eighteenth century, the Paris College of Surgery had nearly 900 students on its books, compared to the Paris medical faculty, which had only 100. Between 1700 and 1789, the estimated numbers of surgeons in France nearly tripled.[10] It is harder to gauge numbers of surgical students in London, where surgical education was often more split between teaching hospitals and private medical schools, but by the early years of the nineteenth century, around 300 students per year were enrolling in medical education in London hospitals, and over 10,000 between 1750 and 1815.[11] A similar pattern of a rising number of surgical students could be seen in the Edinburgh and Glasgow medical schools and university.[12] An important part of surgical training was dissection. Until the mid-eighteenth century, dissection was essentially a demonstration, overlooked by a theatre of enthralled students and even an interested public.[13] As changes in medical training took place, however, dissection became part of the medical education of individual surgeons, who learned their trade for the living upon the corpses of the dead. This was partly based on the need for surgeons to separate themselves from quackery and establish their credentials as authorities on the workings of the body. At the same time, previous emphases laid upon medical theory were being discarded in favour of practical, hands-on knowledge.[14] As one dissection manual put it, there were only two ways to discover the workings of a machine. The first was to be taught by its creator, while the second was 'to take it intirely to pieces'.[15] Since the former was out of the question, the latter was the only recourse. By 1800, Sir Charles Bell was advising young surgeons just embarking on their careers, to begin with a sound theoretical understanding of the body before attempting dissection upon a small corpse, preferably of a young person, whose arteries and blood vessels were more elastic and easier to work with'.[16] Dissection placed a premium on cadavers to be used for experiment.[17] Presumably, it also increased demand for instruments to cut them up.

The apparent growth in demand for instruments offered new opportunities for surgeon's instrument makers. In the mid-eighteenth century, cutlers were the main producers of surgical instruments. There were obvious links between the trades. The skills needed to create sharp

DOI: 10.1057/9781137467485.0010

and durable instruments were common to the manufacture of cutlery, amputation knives and razors. Cutlers, like razor makers, were often innovators in metallurgy, applying their skills to creating sharper but also more elegant and elaborate instruments to appeal to 'gentlemen of the faculty'. In the early to mid-eighteenth century, advertisements by

FIGURE 8 *Trade Card of Henry Patten, Razor Maker, undated, late eighteenth century*
Source: Courtesy of Wellcome Images.

DOI: 10.1057/9781137467485.0010

cutlers sometimes included lancets and other small instruments in their lists of products. Trade cards might also depict surgical instruments while not explicitly mentioning them.

There were signs of specialisation, however, and some saw the manufacture of surgical instruments as a distinct branch. In 1763, the *Universal Director*, a list of London trades, described surgical instrument making as 'a distinct branch from the common Cutlers'.[18] Seventeen names were included, eleven of whom were general manufacturers, five combined instrument and truss making, and one specialised in lancets.[19] In 1800, the prominent cutler, razor maker and instrument maker John Horatio Savigny of London produced a new catalogue of his wide range of surgical instruments. Almost thirty pages long, it was a 'modern compilation' of everything that a surgeon in the 'most eminent hospitals' might need.[20] Amongst the various specialist instruments were 'sets of capital instruments' for cavalry and navy surgeons in the field, elegant pouches and pocket cases in mahogany and tortoiseshell.[21]

In fact, both instrument makers and surgeons experimented with new materials. Cutlers and instrument makers were expected to have some anatomical knowledge to inform the design of their products, as well as the metallurgical skills to create them. In 1800, Robert Bishop, a cutler and surgeon's instrument maker, wrote one of the most comprehensive studies of the importance of steel in surgical instrument manufacture. Bishop argued that cutting instruments were hampered unless the steel was tempered to the exact requirements of usage. If the temper were wrong, the edge would swiftly crumble, leading to skin and tissue damage.[22] Occasionally, instrument makers patented their own devices. In 1800, Savigny submitted a specification for his new 'instrument to be used in surgical operations called a tourniquet', which relied heavily on steel in its construction.[23] Savigny was an expert metallurgist. It is also clear, however, that surgeons contributed to the process of experimentation and innovation. As experts in the body, they were well placed to suggest design changes to instruments based on their own requirements. Specific instruments began to be marketed with something resembling a brand name. In Savigny's catalogue were products such as 'Allanson's double edged knife'.[24] Edward Allanson was a surgeon in Liverpool Infirmary, specialising in amputation and renowned for his invention and use of the instrument.[25] 'Rymer's Tourniquet' was named after its creator, the naval surgeon and medical author James Rymer.[26] The tourniquet was based on Rymer's dissatisfaction with a previous model used

DOI: 10.1057/9781137467485.0010

on board ship, and was put into production.[27] The relationship between surgeon and instrument maker was clearly reciprocal.

How many instruments surgeons actually owned and used is unclear. In some parts of Europe, they relied heavily on sets loaned to them by hospitals.[28] Studies of surgeons in France and Holland suggest that surgeons seldom owned more than a few small, basic instruments.[29] Peter Stanley argues that British surgeons were likely to have owned basic pocket sets of instruments including amputation knives and a saw. A full set would include knives, saws, probes, forceps and lancets, covering the range of most common operations.[30] Newly qualified surgeons, along with established hospital practitioners, were probably the primary market for makers, and were encouraged to equip themselves with good quality kit. The German anatomist Lorenz Heister advocated a pocket set, including lancets for opening veins and abscesses, straight and crooked scissors, forceps, incision knives, probes, a razor and needles.[31] A similar pocket kit was popular with London surgeons, including knives, lancets, scissors and forceps, as well as a bistoury, spatula, quill and plaster box.[32] Dissection called for a similar range of equipment, including 'knives made of the best steel, and as sharp as razors', both curved and straight, and other instruments from probes to scissors and tubes.[33] It was not necessarily the case that numbers of instruments were increasing though. In 1781, John Andree commented upon what he viewed as a 'judicious retrenchment' of the numbers of instruments wielded in the past.[34] For Andree, changes in the material construction of some surgical instruments enabled surgeons to perform operations 'in less time and with more apparent dexterity', and also with less pain and greater chances of a recovery for the patient.[35] Benjamin Bell also cautioned young surgeons to rely on a few trusted instruments rather than confounding themselves under the pressure of the operation with too many choices.[36] Others, however, complained about urban surgeons who, rather than dirty their hands with the messy business of surgery, spent their time inventing new procedures or instruments.[37]

By the 1780s, newspapers began to carry advertisements from cutlers and instrument makers, addressing 'gentlemen of the faculty' specifically puffing their surgical instruments. These advertisements suggest a ready market. Advertising in popular publications presupposed that surgeons were amongst the readership. But it also implied a sufficient customer base for what was, after all, a fairly narrow field of manufacture, to justify the expense of advertising. At the same time that advertisers targeted the

DOI: 10.1057/9781137467485.0010

'genteel' medical faculty, the form of surgical instruments also changed from purely functional to elegant and decorous. Ebony, ivory, tortoiseshell and other exotic materials replaced traditional bone and wooden handles.[38] There were also changes in the ways that surgical instruments were advertised. As with other trades discussed here, surgical instrument makers used polite language and genteel imagery in their advertising to sell something that was essentially prosaic. The trade card of John Chasson, 'razor and surgeon's instrument maker' Chasson's trade card, which is Wellcome Library MS GC EPH 611:8, trade card of John Chasson, undated, late 18th century, shows amputation saws, knives and lancets set against an elegant rococo surround. Chasson laid special emphasis upon the shagreen cases, and the advertisement is repeated in fashionable French. Razor maker Henry Patten's trade card, depicted in Figure 8 displays many instruments, including lancets, which are depicted hanging from a branch of the elaborate surround. Part of this change perhaps relates to broader changes in medicine. Once the poor relations of the medical triumvirate, surgeons were gaining in esteem, prestige and also wealth. Newspaper advertisements followed the form of other types of polite advertising. Instruments upon 'improved plans' by 'celebrated authors' and to the 'highest perfection' could be purchased from Thurgood's Surgeon's-Instrument and Cutlery Manufactory in Fenchurch Street in 1789.[39] In another advertisement, Thurgood adopted the deferential tone of the polite salesman to reassure 'any professional gentleman' that 'nothing shall be wanting on his (Thurgood's) part to render full satisfaction'. He also promoted his own specific 'fibulated sound', which he claimed had already been used by a 'Gentleman of the most respected practice' to save the lives of patients.[40]

Instrument design and surgical technique

It is unsafe to assume that the simple existence or invention of products automatically implies usage.[41] Peter Stanley has also argued that surgery was less about technology than it was about technique.[42] Contemporary evidence from surgeons, however, suggests that the technology of surgical instruments, and in particular cast steel, was an important factor both in design and use. One of the most important elements required by surgeons was fitness for purpose. As James Lucas argued in 1800, it was far better that a surgical instrument was basic and simple in form than of an overcomplicated design. The less the surgeon needed to pay attention

DOI: 10.1057/9781137467485.0010

to the instrument, the better for the patient.[43] The way an instrument was made and the materials of its construction, argued Lucas, greatly affected the technique of the surgeon. A '[t]russ made of steel tempered by rolling [was] much lighter' than those made of iron.[44] A steel blade with the 'cutting edge or sharp point being in order, may facilitate the passage of efficacy' of the knife. Lucas also regarded the ability to carry a high polish as being a sign of perfection in an instrument, and one that could make it easier to gain access to the body.[45] Savigny's catalogue included a page of midwifery tools, including forceps which had 'polished blades' or that were 'polished all over'.[46] Cast steel, as we have seen, could carry a high polish, but it also had other desirable characteristics. Some surgeons insisted that the composition of steel rendered it preferable for specific instruments. The aptly named Samuel Sharp recommended that the canula, a tube for removing fluid, should be made of steel since common silver models were too soft, or became jagged and bruised.[47] He also recommended steel's smooth surface for a 'sound', a probe used for locating stones, and the gorget, a concave cutting knife for removing bladder stones, as well as noting the importance of its elasticity in specula for keeping eyes open during ophthalmic procedures.[48]

Whilst their advertisements were clearly not addressed to patients, instrument makers stressed the utility of their goods in lessening pain and improving the sufferer's lot. Key to this process was the emphasis upon new types of material and manufacture. Incising abscesses and lancing boils were amongst the many running repairs to which a surgeon might attend. Bloodletting was still a mainstay of health regimens, and generally the first option undertaken for treatment. Even in routine procedures like phlebotomy, blunt lancets were more painful to patients. In 1778, for example, Savigny advertised his new lancets to the medical faculty. Recognising that few surgical procedures required as much skill as the opening of a vein, he argued that a keen edge was of paramount importance. Savigny's lancets were 'wrought to such a degree of accuracy, as will greatly lessen the pain of the patient, and totally remove all apprehension of disappointment in the operator'.[49] 'Wrought' suggested the metallurgical processes of temper and fining, in turn emphasising the expertise of the maker. The suggestion of pain reduction for the patient also played to the reluctance of people to submit to the knife. In another advertisement for his 'vertical machine...calculated for the perfecting of Lancets', Savigny was confident that the 'extraordinary degree of accuracy' in their edge would secure the 'approbation of the

DOI: 10.1057/9781137467485.0010

Patient and reputation of the Phlebotomist.[50] It is interesting to note that patients came first. Other surgeons agreed. James Lucas, for one, stressed the importance of 'speedy relief and expeditious ease' for the patient during procedures.[51] Given that patients naturally measured the success of operations by their speed and success, rather than the numbers of individual procedures involved, this was a logical consideration.[52] It is important to note too the role of makers in the repair and maintenance of instruments. As the Edinburgh surgeon Benjamin Bell noted in his advice to trainee surgeons, surgical instruments, and especially lancets, were 'injured' with every use. Given that 'the prevention of pain is with most patients a matter of no small consequence', Bell recommended returning instruments to the cutler to be sharpened after only one or two uses.[53]

In terms of the types of surgical procedures performed, these ranged from routine bodily maintenance like bloodletting, to the repair of accidental injuries, as well as treating specific conditions. Abscesses were a common cause for surgery. If external and visible, they were often left to 'ripen' and fill with fluid, sometimes to huge size, in the belief that the accumulation of pus was a good sign.[54] Once this supposed state of 'maturity' had been reached, the abscess was opened and drained to expose it to the healing properties of air. The site of infection was also important. If the abscess was internal (commonly around the perineum), gaining access was important and required a specific type of instrument. The first stage of the operation was to make a cut and insert a 'director' into the cavity to allow the surgeon access. This was commonly a long, thin, rigid metal instrument. The problem with this approach was that the rigid instrument did not accommodate the curves of the body, making it difficult for surgeons to hold the knife whilst cutting. Samuel Sharp suggested a new curved steel director, with a ring to accommodate the surgeon's thumb. Using the natural tensile qualities of cast steel, it followed the shape of the cavity. The cutting was done with a new type of straight-edged knife, which, according to Sharp, was preferable to curved instruments in this procedure.[55]

Conditions of the eye called for precision tools. In treating cataracts, Benjamin Bell advocated a new small knife in the shape of a 'spear-point lancet', which should be highly polished but also firm and flexible to allow it to penetrate the thick membrane of the cornea. Those surgeons not used to the procedure, cautioned Bell, would find it difficult to perform the operation using the traditional knife.[56] In the accompanying plates,

DOI: 10.1057/9781137467485.0010

Bell noted that the handles should be made of light timber, while 'the steel part of them should be polished in the most exquisite manner' and should be no more than 40 grains in weight. Bell also specified the use of polished steel for trepan saws, hooks and tubes for use in abdominal surgery.[57]

In other cases, there were cosmetic as well as medical considerations for surgery. The treatment of hydrocele, a condition causing a grossly swollen scrotum, was transformed by cast steel. Hydrocele was extremely painful, but also embarrassing for men since it was 'inconvenient from its bulk [and] irksome from its appearance'.[58] Early methods of treatment were simply to incise the scrotum with a lancet and allow the fluid to drain away, or to tap it using a probe or trochar – a small cross-shaped hollow instrument.[59] The problem lay in inserting the canula into the small wound made by the lancet, not least because of the discomfort caused to the patient. Cutting the scrotum with a lancet also risked damaging the testicles and spermatic cords. Tapping the fluid at too early a stage was also considered ineffective, if not sometimes dangerous, since it also risked damage to the testicle, inflammation and fever. Trochars were commonly made of metal, and sometimes silver. John Andree's 'elastic trochar' was 'formed of well-tempered elastic steel'. When inserted into the wound, and with a stilet (a flexible metallic rod) inserted, the elastic properties of steel forced the trochar to open and close, allowing greater control over the flow of fluid.[60] In other procedures, Andree was clear about the need to use the correct instrument with precision. In a description of treatment for phymosis, involving removal of part of the penile foreskin, Andree noted that a curved steel knife was the best instrument, rather than scissors, which bruised as well as cut, causing pain to the patient, 'and [that] no surgeon can be justified giving more pain to the patient than is absolutely necessary'.[61]

But steel was also an important component in changes to major surgical procedures, most notably amputation. Here the need for sharp instruments was obvious. The less time needed to cut through skin, flesh and bone, the better the chances that the patient would survive and recover. Limb amputation put patients at huge risk. First, it involved a massive effusion of blood, placing the body at risk of hypovolemic shock and death. Traumatic shock could result from the harrowing experience of the operation itself, performed without anaesthetic. Third was the risk of secondary infection either introduced from dirty instruments or inherited whilst the wound was healing. Of forty-six amputee patients studied

DOI: 10.1057/9781137467485.0010

by Edward Alanson in 1782, ten died immediately from conditions including lockjaw, haemorrhage of the stump and gangrene; a further eighteen suffered from violent spasms, suppuration and skin loss; and all contracted fevers.[62] For the patient to stand any chance of survival, speed was vital. By the end of the eighteenth century, the strength and edge of cast steel effected changes to the design of several amputation instruments.

The process of amputation was traumatic enough for the patient, but the violence of the operation, and the highly graphic images it produced, also affected practitioners.[63] Once the patient had been brought to theatre and restrained, a metal tourniquet was applied to the limb and tightened to restrict blood flow to the area. As with Savigny's example, the strength and elastic properties of steel were used to maintain firm pressure. Taking hold of the limb, the surgeon commonly made a swift circular cut through the flesh to expose the limb bone. Mid-eighteenth-century amputation knives commonly had curved blades to follow the natural curve of the limb and help the surgeon make a rapid cut. By 1800, however, the preference was for long, narrow and slim knives made of cast steel.[64] Once the flesh had been pared back, the next stage was to sever the limb by cutting through the bone. In young patients, in whom the bone was small, it was possible for the surgeon to cut through the flesh with a knife, but use strong steel nippers (pliers) to quickly cut the bone.[65] Steel saw blades were less likely to snap under the pressure of hacking through bone, while steel-sprung forceps allowed a firmer grip.[66] Benjamin Bell's metacarpal saw was similar in shape to a modern fretsaw, and with cross-cut serrations to minimise splintering. Percival Pott recommended a straight tenon saw for cutting bone, rather than a bow saw as used by Bell.[67] Once this was done, arteries and veins were sutured and, using a new technique, a flap of skin was sewn back over the stump to reduce infection and improve appearance.

Cast steel, then, was an important component in surgical instrument construction. Its strength, elasticity and ability to carry a polish rendered it signally useful in many surgical procedures. Smaller, refined instruments allowed surgeons to operate with more precision, and the use of cast steel was often specified. It has been argued that the greatest changes to the design of surgical instrumentation and, in particular, amputation instruments occurred in the 1830s and 1840s, with the introduction of 'Liston knives', named after the prominent London surgeon Robert

DOI: 10.1057/9781137467485.0010

Liston.[68] The impact of cast steel, however, was already being felt in both the manufacture and use of surgical instruments by 1775.

Conclusion

In the last quarter of the eighteenth century, surgical instruments were part of the milieu of technological – and especially metallurgical – innovation in London that encompassed a range of bodily devices. The properties of cast steel in particular suited the exacting requirements of precision-edged tools. It also allowed changes to existing instruments, from knives to trochars, which, in turn, shaped surgical technique. The relationship between the makers and users of surgical instruments was reciprocal. Makers such as Savigny supplied assorted products tailored to the needs of surgeons, and also serviced and maintained their tools. Surgeons suggested alterations to existing instruments and designed new ones, employing the knowledge of specialist instrument makers to construct prototypes. Nor were the two groups mutually exclusive. Instrument makers like Savigny also clearly had knowledge of anatomy and used it to patent their own instruments. Likewise, some surgeons dabbled in metallurgy and experimented with tempering to refine and modify the tools of their trade. As the numbers of surgical students increased, and dissection became a more prominent component in medical training, so the demand for instruments increased. Surgical instrument makers were able to advertise their products to the medical faculty, and employed the new language of polite advertising to do so. As this occurred, a noticeable change took place, towards a more elegant, rather than purely functional, form in the design of instruments.

But it is also noticeable that the patient was by no means a mere passenger in this process. Surgeons were aware of the fear engendered by knives, as well as the agony of surgery, and took steps to try and mitigate both. Sharper lancets and knives were perceived to cause less pain as well as reducing the length of the operation – a primary concern for patients. Likewise, surgeons looked to minimise collateral damage to surrounding tissue and sought to shape instruments to the specific procedures being undertaken. Although surgical instruments were seldom if ever used for mere cosmetic improvement in this period, they were nonetheless part of a broader enlightenment context of the potential for the material transformation of the bodily fabric.

DOI: 10.1057/9781137467485.0010

Notes

1   A notable exception is Stanley Joel Reiser, *Medicine and the Reign of Technology* (Cambridge: CUP, 1978).

2   Ghislaine Lawrence, 'The Ambiguous Artifact: Surgical Instruments and the Surgical Past' in Christopher Lawrence (ed.), *Medical Theory, Surgical Practice: Studies in the History of Surgery* (London; New York: Routledge, 1992), 295–314.

3   John Kirkup, *The Evolution of Surgical Instruments: An Illustrated History from Ancient Times to the Twentieth Century* (Novato: Jeremy Norman's, 2006); John Kirkup, 'From Flint to Stainless Steel: Observations on Surgical Instrument Composition', *Annals of the Royal College of Surgeons of England*, 75 (1993), pp. 365–374.

4   Elizabeth Bennion, *Antique Medical Instruments* (London: Sotheby Parke Bernet, 1978); Christopher Booth, 'The Rod of Aesculapious: John Haygarth (1740–1827) and Perkins' Metallic Tractors', *Journal of Medical Biography*, 13 (2005), 155–61.

5   Piers Mitchell, *Anatomical Dissection in Enlightenment England and Beyond* (Farnham: Ashgate, 2012).

6   Surgical instruments hardly feature in three standard works, Alison Morrison Low, *Making Scientific Instruments in the Industrial Revolution* (Farnham: Ashgate, 2007); Gloria Clifton and Gerard L'Estrange Turner, *Directory of British Scientific Instrument Makers, 1550–1851* (London: Zwemmer/National Maritime Museum, 1995); R.G.W. Anderson, James Bennett, W.F Ryan (eds), *Making Instruments Count: Essays on Historical Scientific Instruments Presented to Gerard l'Estrange Turner* (Aldershot: Variorum, 1993).

7   Lynda Payne, 'Drain, Blister, Bleed: Surgeons Open the Skin in Georgian England' in Jonathan Reinarz and Kevin Siena (eds), *A Medical History of Skin* (London: Pickering and Chatto, 2013), 24–5.

8   Ibid., 29.

9   Johannes Van Horn, *Micro-Techne; or, a Methodical Introduction to the Art of Chirurgery* (London: printed by J. Darby, 1717), 18.

10 Laurence Brockliss, 'Organization, Training and the Medical Marketplace in the Eighteenth Century' in Peter Elmer (ed.), *The Healing Arts: Health, Disease and Society in Europe, 1500–1800* (Manchester: MUP/Open University, 2004), 356, 364.

11 Susan C. Lawrence, 'Private Enterprise and Public Interests: Medical Education and the Apothecaries' Act, 1780–1825' in Roger French and Andrew Wear (eds), *British Medicine in an Age of Reform* (London: Routledge, 1991), 48; Simon Chaplin, 'Dissection and Display in Eighteenth-Century London' in Piers Mitchell (ed.), *Anatomical Dissection in Enlightenment England and Beyond* (Farnham: Ashgate, 2012), 101.

DOI: 10.1057/9781137467485.0010

12   Helen M. Dingwall, *A History of Scottish Medicine* (Edinburgh: Edinburgh University Press, 2003), 123–4, 127–8; Peter Stanley, *For Fear of Pain: British Surgery, 1790–1850* (Amsterdam and New York: Rodopi, 2003), 161.

13   Chaplin, 'Dissection', 101.

14   Cindy Stelmackowich, 'The Instructive Corpse: Dissection, Anatomical Specimens and Illustration in Nineteenth-Century Medical Education', *Spontaneous Generations: A Journal for the History and Philosophy of Science,* 6:1 (2012), 51–2.

15   G. Thomson, *The Art of Dissecting the Human Body, in a Plain, Easy and Compendious Method* (London: Printed for Joseph Davidson, 1740), iii.

16   Sir Charles Bell, *A System of Dissections, Explaining the Anatomy of the Human Body, The Manner of Displaying the Parts, and Their Varieties in Disease, Volume 1* (Edinburgh: Printed for Mundell and son, 1799), v.

17   Stanley, *For Fear of Pain,* 170.

18   Thomas Mortimer, *The Universal Director or, the Nobleman and Gentleman's True Guide to the Masters and Professors of the Liberal and Polite Arts and Sciences* (London: printed for J. Coote, 1763), 20.

19   Ibid., 21.

20   J.H. Savigny, *A Catalogue of Chirurgical Instruments* (London: Printed by W. Bulmer and Co., 1800), 1.

21   Ibid., 24–5.

22   Robert Bishop, *A Few Observations Relative to Tempered Steel. By R. Bishop, Working Cutler, and Surgeons' Instrument Maker, No. 2, Bruton-Street, New Bond Street* (London: Printed by J. Cundee, 1800), 6.

23   British Library, Patent 2387, Specification of John Horatio Savigny, Surgical Instrument, 31 March 1800.

24   Savigny, *Catalogue,* 9.

25   Robert White, *Practical Surgery: Containing the Description, Causes and Treatment of Each* (London: Printed for T. Cadell, 1796), 168–9.

26   Savigny, *Catalogue,* 9.

27   See James Rymer, *Observations and Remarks Respecting the More Effectual Means of Preservation of Wounded Seamen and Marines on Board of His Majesty's Ships* (London: Printed for J. Donaldson, 1780), 4–6.

28   Lawrence, 'Ambiguous Artifact', 302.

29   Ibid.

30   Stanley, *For Fear of Pain,* 75.

31   Lorenz Heister, *A General System of Surgery in Three Parts. Containing the Doctrine and Management I. Of Wounds, Fractures, Luxations, Tumours, and Ulcers, of All Kinds. II. Of the Several Operations Performed on All Parts of the Body, Volume 1* (London: printed for W. Innys, 1743), 11–12.

32   Samuel Mihles, *The Elements of Surgery. In Which Are Contained All The Essential and Necessary Principles of the Art* (London: Printed for Robert Horsfield, 1764), 3.

DOI: 10.1057/9781137467485.0010

33   Thomson, *The Art*, 6–9.

34   Charles Andree, *Account of an Elastic Trochar, Constructed on a New Principle, for Tapping the Hydrocele, or Watery Rupture* (London: Printed for T. Davis, 1781), 1.

35   Ibid.

36   Benjamin Bell, *A System of Surgery: by Benjamin Bell, Member of the Royal Colleges of Surgeons of Ireland and Edinburgh, Volume 7* (Edinburgh: Printed for Bell and Bradfute, 1796), 298.

37   Payne, 'Drain', 29.

38   John Kirkup, 'From Flint to Stainless Steel: Observations on Surgical Instrument Composition', *Annals of the Royal College of Surgeons of England 75* (1993), 370–1.

39   Advertisement, 'Thurgood's', *Star*, 29 May 1789.

40   Advertisement, 'Surgeon's Instrument and Cutlery Manufactory', *Argus*, 11 April 1789.

41   A point made forcefully in Lawrence, 'Ambiguous Artifacts', 304.

42   Stanley, *For Fear of Pain*, 75.

43   James Lucas, *A Candid Inquiry into the Education, Qualifications, and Offices of a Surgeon-Apothecary* (London: Printed and sold by S. Hazard, 1800), 237.

44   Ibid.

45   Ibid.

46   Savigny, *Catalogue*, 16.

47   Samuel Sharp, *A Treatise on the Operations of Surgery, with a Description and Representation of the Instruments [Sic] Used in Performing Them* (London: Printed for G. Robinson, 1784), 62.

48   Ibid., 103, 106, 159.

49   Advertisement, 'Lancets', *Gazetteer and New Daily Advertiser*, 12 January 1778.

50   Advertisement, 'Lancets', *Daily Advertiser*, 7 November 1776.

51   Lucas, *Candid Inquiry*, 239.

52   Pain, 'Drain', 26.

53   Benjamin Bell, *A System of Surgery. By Benjamin Bell, Member of the Royal College of Surgeons of Edinburgh, Volume 1* (Edinburgh: Printed for Charles Elliott, 1783), 92–3.

54   Ibid., 22.

55   Sharp, *Treatise*, XLIX.

56   Benjamin Bell, *A System of Surgery. By Benjamin Bell, Member of the Royal Colleges of Surgeons of Ireland and Edinburgh, Volume 3* (Edinburgh: printed for Charles Elliot, 1787), 448.

57   Ibid., 533, 524, 533, 536–7.

58   Andree, *Account*, 17.

59   Ibid., 10.

DOI: 10.1057/9781137467485.0010

60　Ibid., 36. A footnote states that the model was made by 'the ingenious Mr Savigny'.

61　John Andree, *Observations on the Theory and Cure of the Venereal Disease* (London: printed for W. Davis, 1779), 70.

62　Edward Alanson, *Practical Observations on Amputation, and the After-Treatment: To Which Is Added, an Account of the Amputation above the Ancle with a Flap* (London: printed for Joseph Johnson, 1782), XIV.

63　For a full description of the operation see Stanley, *For Fear of Pain*, 82.

64　John Kirkup, *A History of Limb Amputation* (London: Springer, 2007), 80.

65　This technique was described in Robert Mynors, *Practical Thoughts on Amputations, &c. By R. Mynors, Surgeon* (Birmingham: Printed by Piercy and Jones, 1783), 86.

66　Ibid.

67　Bennion, *Antique Medical Instruments*, 20–1.

68　See, for example, Stanley, *For Fear of Pain*, 75; Bennion, *Antique Medical Instruments*, 59–60.

DOI: 10.1057/9781137467485.0010

Conclusion: (Re)constructing the Eighteenth-Century Body

Abstract: *This chapter revisits debates about the eighteenth-century body, and the importance of a range of technologies in self-fashioning. As Withey argues throughout the book, the relationship between the body and technology was complex. Cultural shifts in attitudes towards the shaping of the body laid new emphasis upon themes such as 'improvement', politeness and elegance of appearance. As outward features came increasingly to symbolise inner character, so altering the body to create the illusion of harmony was privileged. The physical properties of cast steel allowed makers of a range of products to refine their goods. More than this, it acted as a vector for the articulation of changing ideals of appearance. Cast steel was perhaps the enlightened metal. But, as Withey demonstrates, it was also that with which people had the most regular and intimate contact.*

Withey, Alun. *Technology, Self-Fashioning and Politeness in Eighteenth-Century Britain: Refined Bodies.* Basingstoke: Palgrave Macmillan, 2016.
DOI: 10.1057/9781137467485.0011.

DOI: 10.1057/9781137467485.0011

This book has sought to demonstrate the important relationship between cast steel instruments and the shaping of the human body in the eighteenth century. Cast steel was an enabling material for the refinement of the body. Especially after 1750, the numbers of devices available for bodily transformation proliferated. Part of the reason for this was a cultural shift in attitudes towards expectations of form and appearance. The use of artificial means to alter the fabric of the body, for example, had previously borne connections with vanity and pride. Increasingly, however, the use of bodily technologies became enmeshed not only within narratives of health but also various and changing ideals of body, gender, appearance and form.

First was a new emphasis upon bodily 'improvement', especially in the matter of aesthetic appearance. As David Turner has argued, beauty took on new importance. Bodily features, in tandem with manners and breeding, all contributed to a harmonious whole.[1] As outward features came to symbolise inward characteristics, so attention turned to managing the body and, if necessary, creating the illusion of a harmonious body by disguising flaws and imperfections. Nonetheless, there were debates about vanity and the effeminising effects of fashion upon men in particular. To alter the body for mere fashion was frowned upon.

Second was the important issue, perhaps especially so for men, of control, both over mind and body. This was a period of transition, which began to see debates about gender centre upon the body. Ideals of masculine and feminine behaviour were undergoing changes between 1750 and 1800, not least of which were shifts from politeness towards 'sensibility' and refinement. Some general characteristics remained, however, especially relating to self-control. To master the passions, after all, was to govern the mind. If the mind could be controlled through the conscious exercise of reason, so the body was equally capable of being mastered. Over the course of the eighteenth century, ideas about the mechanistic nature of the body proliferated. As a machine, the human body could be controlled and developed, with faulty parts replaced. To alter the body, whether by plucking the eyebrows or changing its very shape with a postural device, was to control it.

A third key theme was that of the often-ambiguous role of 'nature'. Much attention was paid towards restoring the body to a 'natural' form. It could be argued that the eighteenth-century body, twisted and deformed by all manner of devices such as elastic bandages, neck swings, trusses and collars, was essentially unnatural. Indeed, some contemporaries

DOI: 10.1057/9781137467485.0011

argued that the body in its natural state – that is, unadulterated and unaltered by human intervention – was closest to a bodily ideal. Even the truss maker Timothy Sheldrake cautioned against the arbitrary use of devices and the 'artificial mismanagement' of the body for cosmetic purposes.[2] Debates raged about the moral rectitude of adopting devices to shape the body. Some viewed devices such as steel collars and supports as deceptive, creating an illusion that was soon shattered once they were removed. Nevertheless, the eighteenth century brought about changes to attitudes towards bodily transformation and, in particular, the freedom to 'correct' the vagaries which nature had bestowed upon an individual. Many products were also sold on the promise of restoring the body to a 'natural' form, whether by forcing it into the desired shape or concealing a visible defect. The paradox, therefore, was that people had to use unnatural means to achieve a natural shape.

The vector for so much bodily transformation was steel – a technology that was, itself, also transformed during this period. Especially in London, manufacturers across a wide range of outputs were engaged in processes of development and refinement in metallurgy. Steel was a material that possessed extremely useful physical attributes across diverse trades. But, as part of a wider appreciation of the potential aesthetic appeal of metals (including new substances such as 'Ormolu' and 'Pinchbeck'), it was also a material that was capable of becoming desirable in its own right. The same material that, in the form of jewellery was adorning the necks and clothing of the savant and keeping the time on the fob watch of the beau monde, was also shaving faces, pulling hairs and cutting bodies. To be sure, many other materials were involved in the refinement of the body. Everything from elastic rubber bandages to ivory teeth and wooden legs were also 'technologies of the body'. Steel was one of a plethora of materials used to fashion body form and surfaces. This book does not claim, therefore, that cast steel was the sole driver of change. It was, however, *the* enlightened metal and, arguably, the industrial material with which people came into the most intimate, daily contact.

Much evidence would seem to support the argument that the purposeful transformation of the body, at least for the purposes of meeting new expectations of appearance, was the domain of a fairly narrow section of society – a literate elite, those who attended public events, joined societies, read newspapers and for whom social performance in a public context was important. Nevertheless, those lower down the social scale were also party to broader ideas about bodily transformation. As Jon

DOI: 10.1057/9781137467485.0011

Styles' work on the second-hand market for clothing and accoutrements such as watches in the eighteenth century suggests, people at lower levels of society still wished to participate in fashion, and not simply in order to emulate elites.[3] Clothing was a dialogue, a material expression of values and status. As this book has argued, the same factors that drove people to desire a fashionable appearance extended to the body itself. Questions might be raised about the financial ability of the lower orders to participate in this market, and there a frustrating lack of evidence to suggest any sort of second-hand market for the devices and instruments discussed in this book. Nonetheless, fleeting examples such as the Leicester practitioner noted in Chapter 3, treating 'all classes from all parts of England', hint at a broader willingness across eighteenth-century society to shape the body.

Enlightened bodies?

Notions of bodily appearance were established and communicated through many different channels, including conduct literature to instruct young ladies and gentlemen in polite behaviours, medical texts and self-help books, exercise and dance manuals and even advertising, which appealed to new bodily attributes attainable through the purchase and use of devices and instruments. Bodies that were 'crooked', by contrast, could be viewed pejoratively. Poor posture and gait acted as a barrier to social ambition for both men and women. As the chapter on postural devices suggested, bent bodies and deformed limbs, as well as all manner of other impairments, protuberances and excrescences invited ridicule and caused embarrassment. Many congenital conditions could be troublesome because of their visibility to others. It was difficult to disguise a large inguinal hernia or marked stoop. In many respects, therefore, the body was at the heart of polite rhetoric about social performance.

Traditional narratives of politeness stress the importance of language, gesture, behaviour and deportment. And yet the very form and shape of the human body was also important in conveying the character and civility of the person. Attributes such as symmetry, straightness, proportion and a general elegance or mien, for example, were highly prized and were established through a wide variety of sources. A polite individual had good posture and natural grace in their movement, their neck and spine both straight. As we have seen, a variety of bodily technologies

DOI: 10.1057/9781137467485.0011

were available to alter the physical form of the body. In some cases, it was the physical properties of cast steel that rendered changes to what was materially possible in terms of bodily transformation. In others, however, it was an important component in instruments that were achieving greater prominence in daily grooming routines. As such, steel occupied a unique position as a material that was itself the result of enlightened innovation in metallurgy, but one that could also feed back into a continuum of ideas about the expression of enlightened ideals, including bodily form.

Cast steel rendered changes in the form and structure of posture devices. The makers of stays, trusses, neck swings and other similar devices utilised the springy strength of steel to force the body into the desired shape. Large apparatuses such as neck swings sought to correct spinal deformity, especially in children, which parents viewed as a considerable barrier to social progress later in life. The pain and discomfort reported by the users of such machines underlines the lengths to which people were prepared to go to 'improve' their appearance. Advertising rhetoric played on the use of new materials in their construction, but also emphasised their ability to be concealed about the body. Whilst some attention was paid to the aesthetic appearance of some devices, such as polished steel back supports, such objects were not, of themselves, desirable. Instead they contributed to the creation of a harmonious appearance and were therefore important vectors of politeness.

In many ways, however, it was small, quotidian instruments that proved most important for their part in shaping bodily areas and surfaces, as well as individual features increasingly being regarded as symbols of politeness. Here again, steel was key. The face, for example, was perhaps the most obvious and public of bodily surfaces, and its features were regarded as a window into the character beneath. In general, a clear complexion was prized, even if this meant slathering on layers of creams and powders. Both male and female faces were expected to be smooth; depilation in both sexes was indeed an important component in daily toilette. Cast steel razors gave men a more comfortable means of achieving the new masculine ideal of the smooth face. In turn, they enabled men to reflect the open countenance that symbolised a mind open to new ideas.

Women (and indeed some men) used metal tweezers, often part of a broader range of toilette instruments, to pluck out unsightly facial

DOI: 10.1057/9781137467485.0011

and nasal hairs, as well as shaping the eyebrows – one of many bodily features that were held to reflect inner character. Likewise, a set of clean, unmarked teeth was a visible indicator of the care and attention paid by an individual to their appearance, and involved the use of metal toothpicks of various sorts. Nail 'nippers' and small knives were part of the maintenance of hands, viewed as important indicators of sensibility and character. As the 'principal organs of touch', hands played an equally important role in the public display of the body. Authors such as D. Low and Nicholas Andry set out ideals of shape and proportion, as well as noting the importance of care of the nails. Neat and tidy fingernails, rather than bitten or mangled ones, were again indicators of character. In other cases, the use of steel was one of several factors contributing to the changing significance of some instruments. Spectacles took on new meanings beyond their immediate functionality, coming to represent learning and sagacity, and drawing attention to the 'vision' and gaze of their wearer.

It is important to note, however, that, aside from the discussion of surgical instruments, this book is not centrally concerned with proc-esses that were medical. Indeed, while a great deal of attention has been paid in medical historiography over recent years to the process of 'medicalization', many of the routines and behaviours presented here could be argued to represent a process of 'demedicalization'. They were, however, still centrally concerned with health and hygiene routines and, more importantly, personal routines. From shaving to the wearing of postural devices, spectacles and indeed personal grooming, there was a new focus upon attending to bodily form and appearance. Much of the responsibility for this daily management was shifting towards individu-als. Until the late eighteenth century, barbers (who until 1745 had been barber-surgeons) had been almost solely responsible for the provision of shaving services for men. In this sense, shaving could be considered a 'medical' function performed by practitioners. But as men began to shave themselves, responsibility shifted towards lay individuals. Posture devices and even spectacles were often sold by makers who located themselves outside the medical profession and who sold their products on the basis that people could buy and use them without recourse to a medical practitioner. Even personal grooming routines such as digging out earwax and cutting nails were ultimately matters of bodily health and hygiene, as much as they were about appearance, for which people assumed responsibility. It should also be noted too, though, that the

DOI: 10.1057/9781137467485.0011

period saw increasing specialization by practitioners from oculists to chiropodists, who were claiming dominion over knowledge of specific bodily parts and offering treatment, as well as selling their own devices.

The chronology of change was neither smooth nor linear. The introduction of cast steel neither effected immediate and universal change nor swept aside all that had gone before. Nonetheless, the second half of the eighteenth century was a point of conjunction between technological possibilities and cultural changes. The majority of technologies discussed here were altered, or became more prominent, after 1750. The same period in which cast steel was being introduced across various manufacturing outputs was one that witnessed changes both to the concepts of ideal characteristics of body and gender, and to broader ideas about the body as a conveyor of those ideals. Everything from razors to rupture trusses, tweezers to nail nippers, spectacles to toothpicks were part of the myriad new ways in which people could intervene in the appearance and management of their bodies. Whilst politeness has long been seen more in terms of systems of behaviour, deportment and language, this book has shown how the body itself was capable of being polite. But, in so doing, it has also demonstrated the strong relationship between the body and objects during the Enlightenment, and their centrality to the purposeful fashioning of the body.

Given the often-slim evidence from individual consumers, a sceptic might argue that evidence for the design, production and marketing of products does not indicate widespread use. And yet the weight of evidence in sources from advertising to conduct literature, corporate records, health and medical regimens and even art and literature support both a readiness (and indeed in many cases an expectation) to alter the body, and the means to do so. The quotidian nature of many of the instruments discussed here indeed makes them extremely unlikely to appear in personal testimonies since personal grooming habits, the process of shaving, the wearing of spectacles and so forth were not generally commented upon. This cannot, however, be taken as evidence that they were not carried out. Instead we find the 'ghost' of consumption through the cultural changes in attitudes outlined above, and the technological development of many kinds of instruments and devices. Furthermore, the apparent growth in manufacture, supply and advertising suggests there was a ready market for these goods.

The Enlightenment therefore laid new emphasis upon the body and encouraged individuals to rethink their own appearance and form, as

DOI: 10.1057/9781137467485.0011

well as their manners. It also afforded them new opportunities to do so, through the many products becoming available. Everything from large and visible deformities to unsightly eyebrow hairs, all inhibiting the conveyance of a harmonious self, were able to be controlled, concealed or removed with greater ease as a result of changing technologies. As a key material component across a range of devices, instruments and machines, cast steel was at the very heart of this process.

Notes

1 David M. Turner, 'The Body Beautiful' in Carole Reeves (ed.), *A Cultural History of the Human Body in the Enlightenment* (London: Bloomsbury, 2010), 114–15.
2 David Turner and Alun Withey, 'Technologies of the Body': Polite Consumption and the Correction of Deformity in Eighteenth-Century England', *History*, 99:338 (2014), 788.
3 Jon Styles, *The Dress of the People: Everyday Fashion in Eighteenth-Century England* (New Haven, CT: Yale University Press, 2008).

DOI: 10.1057/9781137467485.0011

Select Bibliography

Primary Sources

(a) Manuscript Sources and Ephemera

Bedfordshire and Luton Archives:
MS L30/13/12/12, Letter from Annabel, lady Lucas to Mary Jemima Robinson, exact date unknown, 1774.
MSS QSR/18/1801/126,142,151,154, Gaoler's general bills for county gaol, various dates, c. 1801.

Birmingham Archives:
Matthew Boulton papers, MS 3782/12/23/33, Letter from William Allen (London) to MB, 2 April 1764.

The British Library :
551.a.32, A Collection of Medical Advertisements, Seventeenth and Eighteenth Century.
C.112.f9, A Collection of Seventeenth Century Medical Advertisements.
Add. MS 4805, Letter Henry St. John, 1st Viscount Bolingbroke to Jonathan Swift, 27 September 1729.
Patent number 3458, 15 June 1811, Specification for obtaining motive power.
Patent no. 1211, 1779, 'Specification for Machines &C for Gymnastick Exercises.
Patent Number 1716, Specification of John Horatio Savigny, 8 December 1789.
Patent Number 2387, Specification of John Horatio Savigny, 31 March 1800.

DOI: 10.1057/9781137467485.0012

Patent Number 1458, Specification of John Henry Savigny, 4 December 1784.

British Optical Association Museum:
Collections of Trade Material and Advertisements.

Cumbria Record Office:
MS WQ/SR/403/14, Petition of James Noble to Quarter Sessions at Appleby, Michaelmas Sessions, 1777.

Gwynedd Archives:
MSS ZQS/E1814/6, Meirionydd county treasurer's accounts, 17 April 1814.

Herefordshire Record Office:
MS BG11/17/5/72, Details of Charges against Edmund Hawley, 1686.

Lancashire Archives:
MS QSP/905/7, Quarter sessions records 1703/4.

London Metropolitan Archives:
SC/GL/TCC, Trade Card Collection.

Manchester Archives:
MS L24/1 (Box 24), 'The Prices of Lancashire Tools &c Manufactured by Peter Stubs, Warrington', undated, late eighteenth century.

National Library of Wales
MS SD/1693/208, Probate inventory of William Lloyd of Carmarthen, January 1693.

Nottinghamshire Archives:
MS DDE 98/181, Receipt for monies due from 'Mr Evetts', Month unclear, 1772.
MS DDE 98/192, Receipt for monies due from 'Mr Evetts to Samuel Clowes, 26 August 1773.
MS DDE 98/105, Receipt for monies due from 'Mr Evetts' to William Sharp, 23 January 1769.

Oxford University, Bodleian Library:
John Johnson Collection of Printed Ephemera.

DOI: 10.1057/9781137467485.0012

Douce Portfolio 139 (99), 'Fine Ground Spectacles for All Sights', undated.

Science Museum:
MS 1951–685/88, Trade Card for John Yarwell, Optician, c. 1697.
MS 1951–785/63, Trade card of Thomas Ribright, c. 1753–72.
MS 1934-0096, Trade card of John Gilbert, undated, early nineteenth century.

Victoria and Albert Museum:
MS E10–98 and E36–98, Trade catalogue of Ross and Co. Ironmongers, c. 1797.
MS E126/96, Trade catalogue of Ernst and Co. Ltd, 1811.

Warwickshire Record Office:
MS L6/1320, Receipt bill from William Orchard to George Lucy esq., 11 May 1766.

(b)   Electronic Sources

Dictionary of Traded Goods and Commodities < www.british-history.ac.uk/no-series/traded-goods-dictionary/1550–1820>
Electronic Enlightenment, ed. McNamee, Robert et al, Version 2.2, University of Oxford, 2001 <www.e-enlightenment.com>
Gale British Newspapers 1600–1950.
Letters from Lord Chesterfield to his Son <www.gutenberg.org/files/3361/3361-h/3361-h.htm>
London Lives, 1690–1800 <www.londonlives.org>
Proceedings of the Old Bailey, 1674–1913, <www.oldbaileyonline.org>

(c)   Printed Primary Sources

(1)   *Newspapers and Periodicals*

Adams' Weekly Courant.
The Argus.
Bath Chronicle.
Caledonian Mercury.
Critical Review or Annals of Literature.
Courier.
The Courier and Evening Gazette.

DOI: 10.1057/9781137467485.0012

Daily Advertiser.
Daily Courant.
The Daily Post.
The Derby Mercury.
The Dublin Mercury.
Gazetteer and New Daily Advertiser.
General Advertiser and Morning Intelligencer.
The Leeds Intelligencer.
Lincoln, Rutland and Stamford Mercury.
Lloyd's Evening Post.
London Courant and Westminster Chronicle.
London Daily Advertiser.
London Daily Post.
London Evening Post.
London Journal.
Mist's Weekly Journal.
The Morning Chronicle.
The Morning Herald and Daily Advertiser.
The Morning Post.
Morning Post and Daily Advertiser.
Newcastle General Magazine.
The Oracle.
The Oracle and Public Advertiser.
Oxford Magazine or University Museum.
Penny London Post or The Morning Advertiser.
Post Boy.
Post Man and the Historical Account.
The Public Advertiser.
The Public Ledger.
Read's Weekly Journal or British Gazetteer.
Reading Mercury and Oxford Gazette.
St James Chronicle or the British Evening Post.
The Scots Magazine.
The Star.
Tattler.
True Briton.
The Universal Magazine.
Weekly Journal or British Gazetteer.
Weekly Journal with Fresh Advices Foreign and Domestick.

DOI: 10.1057/9781137467485.0012

The Westminster Journal.
Whitehall Evening Post or London Intelligencer.
The World.
World and Fashionable Advertiser.

(2) Other Printed Sources

Alanson, Edward, *Practical Observations on Amputation, and the After-Treatment:* (London: 1782).

Allen, John, *Dr Allen's Synopsis Medicine* (London: 1730).

Andree, Charles, *Account of an Elastic Trochar, Constructed on a New Principle, for Tapping the Hydrocele, or Watery Rupture* (London: 1781).

Andree, John, *Observations on the Theory and Cure of the Venereal Disease* (London, 1779).

Andry, Nicholas, *Orthopædia: or, the Art of Correcting and Preventing Deformities in Children* (London:1743).

Anon, *Alfred Unmasked or the New Cataline; Intended as a Pair of Spectacles for the Short-Sighted Politicians of 1789* (London: 1789).

Anon, *Aristotle's Compleat Master Piece. In Three Parts; Displaying the Secrets of Nature in the Generation of Man* (London: 1749).

Anon, *Aristotle's New Book of Problems, Set Forth by Way of Question and Answer.* (London: 1725).

Anon, *The Art of Nursing: or the Method of Bringing up Young Children According to the Rules of Physick for the Preservation of Health and Prolonging Life* (London: 1733).

Anon, *Beatrice, or the Inconstant... Volume 1*(London: 1788).

Anon, *By Permission of the Worshipful the Mayor, Now Exhibiting at Miss Newton's, Milliner, the Powers of Imagination or the Senses Deceived* (Hereford: 1795).

Anon, *A Catalogue of Optical, Philosophical and Mathematical Instruments Made and Sold by Henry Pyefinch at the Golden Quadrant, Sun and Spectacles* (London: 1765).

Anon, *Eliza: Or the History of Miss Granville* (London: 1766).

Anon, *A Full Account of Mr John Harris, the English Hermit* (Banbury: 1800).

Anon, *The Gentleman's Companion to the Toilet or a Treatise on Shaving, by a London Hair Dresser* (London: 1844).

Anon, *The History of Miss Harriott Fitzroy and Miss Emilia Spencer* (London: 1767).

Anon, *The Ladies Library or Encyclopaedia of Female Knowledge* (London: 1790).

DOI: 10.1057/9781137467485.0012

Anon, 'Miscellania', *Review of the State of the British Nation* (London: 1710).

Anon, *A Pair of Spectacles for Short-sighted Politicians* (London: 1765).

Anon, *The Pantheonites. A Dramatic Entertainment as Performed at the Theatre-Royal in Hay-Market* (London: 1773).

Anon, *The Polite Academy, or School of Behaviour for Young Gentlemen and Ladies* (London: 1762).

Anon, *Village Memoirs: In a Series of Letters Between a Clergyman and his Family in the Country, and his Son in Town* (London: 1765).

Ayscough, James, *A Short Account of the Eye and Nature of Vision* ... (London: 1754).

Beckett, William, *Practical Surgery Illustrated and Improved* (London: 1740).

Bell, Benjamin, *A system of surgery. By Benjamin Bell, Member of the Royal College of Surgeons of Edinburgh, 7 Volumes* (Edinburgh: 1783–96).

Bell, Sir Charles, *A System Of Dissections, Explaining the Anatomy of the Human Body, the Manner of Displaying the Parts, and Their Varieties in Disease, Volume 1* (Edinburgh: 1799).

Bennett, William, *A Dissertation on the Teeth and Gums, and the Several Disorders to Which They Are Liable* (London: 1779).

Bigg, Henry Heather, *The Gentle Treatment of Spinal Curvature* (London: 1875).

Bishop, Robert, *A Few Observations Relative To Tempered Steel. By R. Bishop, Working Cutler, and Surgeons" Instrument Maker, No. 2, Bruton-Street, New Bond Street* (London: 1800).

Brand, Timothy, *Chirurgical Essays on the Cure of Ruptures and the Pernicious Consequences of Referring Patients to Truss Makers* (London: 1785).

Bu'choz, Pierre-Joseph, *The Toilet of Flora* ... (London: 1772).

Bulwer, John, *Anthropometamorphosis or Man Transform'd or the Artificial Changeling* ... (London: 1653).

Caulfield, James, *Blackguardiana: or A Dictionary Of Rogues, Bawds, Pimps, Whores, Pickpockets, Shoplifters* ... (London: 1793).

Chambers, Amelia, *The Ladies Best Companion* ... (London: 1775).

Cheshire, John, *The Gouty Man's Companion; or a Dietetical and Medicinal Regime* (Nottingham: 1747).

Churchill, Charles, *Night, An Epistle to Robert Lloyd* (London: 1761).

Cleland, John, *Institutes of Health* (London: 1761).

Clubbe, John, *A Letter of Free Advice to a Young Clergyman* (London: 1765).

DOI: 10.1057/9781137467485.0012

Collings, Samuel, '*Carlo Khan's Triumphant Entry into St. Stephen's Chapel*' (London: 1784).

Colman, George, *The Connoisseur by Mr. Town, Critic and Censor General... Volume 2* (London: 1757).

Cooke, James, *Mellificium chirurgiæ: or, the Marrow Of Chirurgery* (London: 1717).

Coriat Junior, *Another Traveller or Tritical Observations Made on a Journey Through the Netherlands* (London: 1766).

Cruickshank, Isaac, *Frith the Madman Hurling Treason at the King* (London: 1790).

Curtis, Richard, *A Treatise on the Structure and Formation of the Teeth...* (London: 1769).

Darly, Matthew, *The Optical Contrast* (London: 1771).

Darwin, Erasmus, *A Plan for the Conduct of Female Education, in Boarding Schools, Private Families, and Public Seminaries. By Erasmus Darwin, M.D. F.R.S* (Philadelphia: 1798).

Dulaure, J.A., *Pogonologia: or a Philosophical and Historical Essay on Beards, translated from the French* (Exeter: 1789).

Faraday, Michael, 'An Analysis of Wootz, or Indian Steel. By M. Faraday, Chemical Assistant to the Royal Institution, *Quarterly Journal of Science*, VII (1819), 288–290.

Fuller, Francis, *Medicina Gymnastica; or Every Man His Own Physician* (London: 1777).

'Gentleman, Private', *New Inventions and New Directions, Productive of Happiness to the Ruptured* (London: 1800).

Gray, Thomas, *An Elegy Wrote in a Country Churchyard* (London: 1751).

Hartley, David, *Observations on Man, His Frame, His Duty, and His Expectations. In two parts, Volume 1* (London: 1749).

Hay, William, *Deformity: An Essay* (London: 1754).

Haywood, Eliza, *A Present for a Servant-Maid. Or the Sure Means of Gaining Love and Esteem* (London: 1744).

Heister, Lorenz, *A General System of Surgery in Three Parts. Containing the Doctrine and Management I. Of Wounds, Fractures, Luxations, Tumours, and Ulcers, of All Kinds. II. Of The Several Operations Performed on All Parts of the Body, Volume 1* (London: 1743).

Hill, John, *The Conduct of a Married Life: Laid Down in a Series of Letters Written by the Honourable Julia-Susannah Seymour, to a Young Lady...* (London: 1754).

DOI: 10.1057/9781137467485.0012

Hogarth, William, *The Analysis of Beauty: Written With a View of Fixing the Fluctuating Ideas of Taste* (London: 1753).

Jones, Philip, *An Essay on Crookedness, or Distortions of the Spine* (London: 1788).

Juillon, Paul Eurialius, *A Practical Essay on the Human Teeth...* (London: 1781).

Langford, Abraham, *A Catalogue of the Genuine Stock in Trade of Mr Stephen Quillet, Jeweller and Goldsmith* (London: 1751).

Le Maitre, Michael, *Advice on the Teeth: With Some Observations And Remarks* (London: 1782).

Love, James, *All Sorts of Italian, French and English Perfumes and Powders, with a Variety of Choice and Curious Articles* (London: 1800).

Low, D., *Chiropodologia, or A Scientific Enquiry into the Causes of Corns, Warts, onions and other Painful or Offensive Cutaneous Excrescences...* (London: 1785).

Lucas, James, *A Candid Inquiry into the Education, Qualifications, and Offices of a Surgeon-Apothecary* (London: 1800).

McKittrick Adair, James, *An Essay on Regimen, for the Preservation of Health, Especially of the Indolent, Studious, Delicate and Invalid* (London: 1799).

Manning, Henry, *Modern Improvements in the Practice of Surgery* (London: 1780).

Martin, Benjamin, *An Essay on Visual Glasses (Commonly Called Spectacles)...* (London: 1758).

Merande, D., *A Succinct Account of a Machine, Newly Invented for the Cure of Praeternatural Curvatures of the Spine* (London: 1768).

Mihles, Samuel, *The Elements of Surgery. In Which Are Contained All the Essential and Necessary Principles of the Art; With an Account of the Nature and Treatment of Chirurgical Disorders, and a Description of the Operations, Bandages, Instruments, and Dressings, According to the Modern and Most Approved Practice* (London: 1764).

Moises, Hugh, *An Appendage to the Toilet: Or an Essay on the Management of the Teeth. Dedicated to the Ladies* (London: 1798).

Mortimer, Thomas, *The Universal Director or, the Nobleman and Gentleman's True Guide to the Masters and Professors of the Liberal and Polite Arts and Sciences* (London: 1763).

Mynors, Robert, *Practical Thoughts on Amputations, &c. By R. Mynors, surgeon* (Birmingham: 1783).

Nelson, James, *An Essay on the Government of Children* (London: 1756).

DOI: 10.1057/9781137467485.0012

Nicholson, William, 'Philosophical Discquitions (sic) on the Processes of Common Life: - Art of Shaving', *A Journal of Natural Philosophy, Chemistry and the Arts*, Volume 1 (1802), 47–51.

'P', 'Joe Barrington: Tub Village Barber', *The Olio: Museum of Entertainment*, Volume 2, (London: 1804), 7–9.

Repton, John Adey, *Some Account of the Beard and the Moustachio, Chiefly from the Sixteenth to the Eighteenth Century* (London:1839).

Rhodes, E., *Essay on the Manufacture, Choice and Management of a Razor by E. Rhodes, Cutler of Sheffield* (Sheffield: 1824).

Rowlandson, Thomas, *Barber Woodward* (London: 1799).

Rowley, William, *A Treatise on One Hundred and Eighteen Principle Diseases of the Eyes and Eyelids* (London: 1790).

Rymer, James, *Observations and Remarks Respecting the More Effectual Means of Preservation of Wounded Seamen and Marines on Board of His Majesty's Ships* (London: 1780).

Savigny, John Horatio, *A Catalogue of Chirurgical Instruments* (London: 1800).

Savigny, John Horatio, *Treatise on the Use and Management of a Razor: with Practical Directions Relative to Its Appendages* (London: 1776).

Sharp, Samuel, *A Treatise on the Operations of Surgery, with a Description and Representation of the Inrstruments [Sic] Used in Performing Them* (London: 1784).

Sheldrake, Timothy, *Observations of the Causes of Distortions of the Legs of Children* (London: 1794).

Sheldrake, Timothy, *An Essay on the Various Causes of the Distorted Spine* (London: 1783).

Spilsbury, Francis, *Every Lady And Gentleman Their Own Dentist, As Far As the Operations Will Allow...* (London: 1791).

Stodart, James, 'On an Experiment to Imitate the Damascus Sword Blade', *Journal of Natural Philosophy, Chemistry and the Arts*, VII (1804), 231–2.

Stodart, James and Faraday, Michael, 'Experiments on the Alloys of Steel, Made with a View to Its Improvement', *Quarterly Journal of Science, Literature and the Arts*, 9 (1820), 319–330.

Thomson, G., M.D., *The Art of Dissecting the Human Body, in a Plain, Easy and Compendious Method* (London: 1740).

Thornton, Robert, *Medical Extracts. On the Nature of Health, with Practical Observations: And the Laws of the Nervous and Fibrous Systems* (London: 1795).

DOI: 10.1057/9781137467485.0012

Tissot, Samuel, *Three Essays: First on the Disorders of People of Fashion, Translated by Francis Bacon* (Dublin: 1772).

Tweedie, James, *Hints on Temperance and Exercise, Shewing their Advantage in the Cure of Dyspepsia, Rheumatism, Polysarcia and Certain Stages of Palsy* (London: 1799).

Van Horn, Johannes, *Micro-Techne; or, a Methodical Introduction to the Art of Chirurgery* (London: 1717).

Weaver, John, *Anatomical and Mechanical Lectures upon Dancing. Wherein Rules and Institutions for That Art Are Laid Down and Demonstrated* (London: 1721).

White, Robert, M.D., *Practical Surgery: Containing the Description, Causes and Treatment of Each* (London: 1796).

Wooffendale, Robert, *Practical Observations on the Human Teeth by Robert Wooffendale, Surgeon-Dentist, Liverpool* (London: 1783).

Secondary Sources

Books

Allen, Julia, *Swimming with Dr Johnson and Mrs Thrale: Sport, Health and Exercise in Eighteenth-Century England* (Cambridge: Lutterworth, 2012).

Anderson, R.G.W, Bennett, James, Ryan, W.F. (eds), *Making Instruments Count: Essays on Historical Scientific Instruments Presented to Gerard l'Estrange Turner* (Aldershot: Variorum, 1993).

Barker-Benfield, G.J., *The Culture of Sensibility: Sex and Society in Eighteenth-Century Britain* (Chicago: University of Chicago Press, 1992).

Barraclough, Kenneth, *Steelmaking before Bessemer Volume 2: Crucible Steel – the Growth of Technology* (London: IOM, 1984).

Barty-King, Hugh, *Eyes Right: The Story of Dollond and Aitchison, 1750–1985* (London: Quiller, 1986).

Bennion, Elizabeth, *Antique Medical Instruments* (London: Sotheby Parke Bernet, 1978).

Beresford, John (ed.), *The Diary of a Country Parson: The Reverend James Woodforde, Vol. V, 1797–1802* (Oxford: Clarendon Press, 1968 edition).

Berg, Maxine and Eger, Elizabeth, *Luxury in the Eighteenth Century: Debates, Desires and Delectable Goods* (Basingstoke: Palgrave, 2003).

DOI: 10.1057/9781137467485.0012

Berry, Helen, *Gender, Society and Print Culture in Late Stuart England: The Cultural World of the 'Athenian Mercury'* (Aldershot: Ashgate: 2003).

Biagioli, *Galileo's Instruments of Credit: Telescopes, Images, Secrecy* (Chicago: University of Chicago Press, 2007).

Brewer, John, *The Pleasures of the Imagination: English Culture in the Eighteenth Century* (London: Harper Collins, 1997).

Brewer, John and Porter, Roy (eds), *Consumption and the World of Goods* (London: Routledge, 1994).

Brooke, Jon Hedley, *Science and Religion: Some Historical Perspectives* (Cambridge: CUP, 1991).

Brown, *Foul Bodies: Cleanliness in Early America* (New Haven, CT: Yale University Press, 2009).

Campkin, Ben and Cox, Rosie, *Dirt: New Geographies of Cleanliness and Contamination* (London: Tauris, 2012).

Carter, Philip, *Men and the Emergence of Polite Society, 1660–1800* (London: Routledge, 2000).

Cavallo, Sandra, *Artisans of the Body in Early Modern Italy: Identities, Families and Masculinities* (Manchester: Manchester University Press, 2007).

Clifton, Gloria and L'Estrange Turner, Gerard, *Directory of British Scientific Instrument Makers, 1550–1851* (London: Zwemmer/National Maritime Museum, 1995).

Cohen, Michèle, *Fashioning Masculinity: National Identity and Language in the Eighteenth Century* (London and New York: Routledge, 1996).

Corson, Richard, *Fashions in Eyeglasses from the Fourteenth Century to the Present Day* (London: Peter Owen, 1967).

Corson, Richard, *Fashions in Makeup* (London: Peter Owen, 1972).

Cosgrove, Denis and Daniels, Stephen (eds), *The Iconography of Landscape: Essays on the Symbolic Representation, Design and Use of Past Environments* (Cambridge: CUP, 1988).

D'Allemagne, Henry Rene, *Decorative Antique Ironwork: A Pictorial Treasury* (New York: Dover Publications, 1968).

Davidoff, Leonore and Hall, Catherine (eds), *Family Fortunes: Men and Women of the English Middle Class, 1780–1850* (London: Routledge, 1987).

Dingwall, Helen, *A History of Scottish Medicine* (Edinburgh: Edinburgh University Press, 2003).

Donald, Diana, *The Age of Caricature: Satirical Prints in the Reign of George III* (New Haven and London: Yale University Press, 1996).

DOI: 10.1057/9781137467485.0012

Duffy, Christopher, *The Military Experience in the Age of Reason* (London: Routledge & Kegan Paul, 1987).

Evans, Chris and Ryden, Goran, *Baltic Iron in the Atlantic World in the Eighteenth Century* (Leiden: Brill, 2007).

Feher, Michael, Nadaff, Ramona and Tazi, Nadia (eds), *Fragments for a History of the Human Body, Volume Two* (New York: Zone, 1990).

Findlen, Paula (ed.), *Early Modern Things: Objects and Their Histories, 1500–1800* (London: Routledge, 2013).

Fox, Celina, *The Arts of Industry in the Age of Enlightenment* (New Haven, CT: Yale University Press, 2010).

Gilman, Sander, *Picturing Health and Illness: Images of Identity and Difference* (Baltimore: Johns Hopkins Press, 1995).

Golinski, Jan, *Science as a Public Culture: Chemistry and Enlightenment in Britain, 1760–1820* (Cambridge: Cambridge University Press, 1992).

Goring, *The Rhetoric of Sensibility in Eighteenth-Century Culture* (Cambridge: CUP, 2004).

Hadfield, Robert, *Faraday and his Metallurgical Researches, With Special Reference to Their Bearing on the Development of Alloy Steels ...* (London: Chapman and Hall, 1931).

Harvey, Karen, *Reading Sex in the Eighteenth Century: Bodies and Gender in English Erotic Culture* (Cambridge: Cambridge University Press, 2004).

Harvey, *The Little Republic: Masculinity and Domestic Authority in Eighteenth-Century Britain* (Oxford: OUP, 2014).

Hitchcock, Tim, *English Sexualities, 1700–1800* (Basingstoke: Macmillan, 1997).

Hitchcock, Tim and Cohen, Michèle, *English Masculinities, 1660–1800* (London: Routledge, 1999).

Jones, Colin, *The Smile Revolution in Eighteenth Century Paris* (Oxford: OUP, 2014).

Kemp, Martin, *Visualisations: The Nature Book of Art and Science* (Oxford: OUP, 2000).

Kirkup, John, *The Evolution of Surgical Instruments: An Illustrated History from Ancient Times to the Twentieth Century* (Novato: Jeremy Norman's: 2006).

Lamb, Jonathan, *The Evolution of Sympathy in the Long Eighteenth Century* (London: Pickering and Chatto, 2009).

Laqueur, Thomas, *Making Sex: Body and Gender from the Greeks to Freud* (Cambridge, MA: Harvard University Press, 1990).

DOI: 10.1057/9781137467485.0012

Lindberg , David C. and Numbers, Ronald L. (eds), *God and Nature: Historical Essays on the Encounter between Christianity and Science* (Berkeley: University of California Press, 1986).

Lupton, *Medicine as Culture: Illness, Disease and the Body* (London: Sage, 2012 edition).

Lysons, Daniel, *The Environs of London,* Volume 1 (London: T. Cadell and W. Davies, 1891).

McConnell, Anita, *Jesse Ramsden (1735–1800): London's Leading Scientific Instrument Maker* (Aldershot: Ashgate, 2007).

Martin, Morag, *Selling Beauty: Cosmetics, Commerce and French Society, 1750–1830* (Baltimore: Johns Hopkins Press, 2009).

Millburn, John, *Adams of Fleet Street: Instrument Makers to King George III* (Aldershot: Ashgate, 2000).

Morrison-Low, Alison, *Making Scientific Instruments in the Industrial Revolution* (Aldershot: Ashgate, 2007).

Outram, Dorinda, *The Englightenment* (Cambridge: CUP, 1995).

Pelling, Margaret, *The Common Lot: Sickness, Medical Occupations and the Urban Poor in Early Modern England* (London: Longman, 1998).

Pointon, Marcia, *Brilliant Effects: A Cultural History of Gemstones and Jewellery* (New Haven, CT: Yale University Press/Paul Mellon Centre for Studies in British Art, 2010).

Polanyi, Michael, *The Study of Man* (Chicago: University of Chicago Press, 1966 edition).

Porter, Roy, *Flesh in the Age of Reason* (London: Penguin, 2003).

Porter, Roy, *Quacks: Fakers and Charlatans in Medicine* (Stroud: Tempus, 2001).

Raphael, D.D. and Macfie, A.L. (eds), Adam Smith, *The Theory of Moral Sentiments* (Oxford: Oxford University Press, 1976).

Reiser, Joel Stanley, *Medicine and the Reign of Technology* (Cambridge: CUP, 1978).

Roberts, Lissa L., Schaffer, Simon and Dear, Peter (eds), *The Mindful Hand: Inquiry and Invention from the Late Renaissance to Early Industrialisation* (Chicago: University of Chicago Press, 2008).

Robinson, Nicholas K., *Edmund Burke: A Life in Caricature* (New Haven, CT: Yale University press, 1996).

Roper, Michael and Tosh, John (ed.), *Manful Assertions: Masculinities in Britain Since 1800* (London: Routledge, 1991).

Rosenthal, J. William, *Spectacles and Other Vision Aids: A History and Guide to Collecting* (San Francisco: Normal Books, 1996).

DOI: 10.1057/9781137467485.0012

Ruhrah, John (ed.), *Pediatrics of the Past* (New York: Paul B. Hoeber, 1925).

Semple, Janet, *Bentham's Prison: A Study of the Panopticon Penitentiary* (Oxford: Clarendon Press, 1993)

Sennett, Richard, *The Fall of Public Man* (London: Faber and Faber, 1976).

Shapin, Steven and Schaffer, Simon, *Leviathan and the Air Pump: Hobbes, Boyle and the Experimental Life* (Princeton: Princeton University Press, 1995).

Shepherd, Alexandra, *Meanings of Manhood in Early Modern England* (Oxford: OUP, 2008).

Smith, Virgina, *Clean: A History of Personal Hygiene and Purity* (Oxford: OUP, 2007).

Sorge-English, Lynn, *Stays and Body Image in London: The Staymaking Trade, 1680–1810* (London: Pickering and Chatto, 2011).

Stafford, Barbara, *Imaging the Unseen in Enlightenment Art and Medicine* (Cambridge, MA: MIT Press, 1991).

Stanley, Peter, *For Fear of Pain: British Surgery, 1790–1850* (Amsterdam and New York: Rodopi, 2003).

Stewart, Larry, *The Rise of Public Science: Rhetoric, Technology and Natural Philosophy in Newtonian Britain, 1660–1750* (Cambridge: Cambridge University Press, 1992).

Styles, John, *The Dress of the People: Everyday Fashion in Eighteenth-Century England* (New Haven, CT: Yale University Press, 2007).

Timmerman, Carsten and Anderson, Julie (eds), *Devices and Designs: Medical Technologies in Historical Perspective* (London: Palgrave Macmillan, 2006).

Tosh, John, *Masculinities in Britain since 1800* (London: Routledge, 1991).

Tosh, John, *Manliness and Masculinities in Nineteenth-Century Britain* (London: Pearson, 2005).

Turner, Bryan S., *The Body and Society,* second edition (London: Sage Publications, 1996).

Turner, David, M., *Disability in Eighteenth-Century England: Imagining Physical Impairment* (London: Routledge, 2012).

Van Sant, A.J., *Eighteenth-Century Sensibility and the Novel: The Senses in Social Context* (Cambridge: CUP, 2004).

Vickery, Amanda, *Behind Closed Doors: At Home in Georgian England* New Haven and London: Yale University Press, 2009).

Vigarello, George, *Histoire Du Corps Volume 1* (Paris: Seuil, 2006).

DOI: 10.1057/9781137467485.0012

Vila, Anne C. (ed.), *A Cultural History of the Senses in the Age of Enlightenment* (London: Bloomsbury, 2014).

von Mallinckrodt, Rebekka and Schattner, Angela (eds), *Sport and Physical Exercise in Early Modern Culture* (Farnham: Ashgate, 2016).

Voskuhl, Adelheid, *Androids in the Enlightenment: Mechanics, Artisans and Cultures of the Self* (Chicago: University of Chicago Press, 2013).

Warhman, Dror, *The Making of the Modern Self: Identity and Culture in Eighteenth Century England* (New Haven, CT: Yale University Press, 2007).

Wiesner, Merry E., *Women and Gender in Early Modern Europe, Second Edition* (Cambridge: CUP, 2000).

Willach, Ralf, *The Long Route for the Invention of the Telescope* (Philadelphia: American Philosophical Society, 2008).

Winkler, Wolf, *A Spectacle of Spectacles: Exhibition Catalogue* (Edition Leipzig, 1988).

Chapters in Edited Collections

Auerbach, Emily, 'An Excellent Heart: Sense and Sensibility' in Bloom, Harold (ed.) *Jane Austen* (London: Chelsea House, 2009), 251–78.

Bennett, James A., 'Shopping for Instruments in London and Paris' in Smith, Pamela H. and Findlen, Paula (eds), *Merchants and Marvels: Commerce, Science and Art in Early Modern Europe* (London: Routledge, 2002), 370–98.

Berg, Maxine, 'New Commodities, Luxuries and Their Consumers in Eighteenth-Century England' in Berg, Maxine and Clifford, Helen (eds), *Consumers and Luxury* (Manchester: Manchester University Press, 1999), 63–87.

Brockliss, Lawrence, 'Organization, Training and the Medical Marketplace in the Eighteenth Century' in Peter Elmer (ed.), *The Healing Arts: Heath, Disease and Society in Europe, 1500–1800* (Manchester: MUP/Open University, 2004), 346–51.

Chaplin, Simon, 'Dissection and Display in Eighteenth-Century London' in Piers Mitchell (ed.), *Anatomical Dissection in Enlightenment England and Beyond* (Farnham: Ashgate, 2012), 95–114.

Davis, Lennard J., 'Dr Johnson, Amelia and the Discourse of Disability in the Eighteenth Century' in Deutsch, Helen and Nussbaum, Felicity (eds), *'Defects': Engendering the Modern Body* (Ann Arbor: University of Michigan Press, 2000), 54–74.

DOI: 10.1057/9781137467485.0012

Fitzmaurice, Susan, 'Changes in the Meanings of *Politeness* in Eighteenth Century England' in Culpeper, Jonathan and Kadar, Daniel Z. (eds), *Historical (Im)Politeness* (Bern: Peter Lang, 2010), 87–116.

Hargreaves, Anne, 'Dentistry in the British Isles' in Hillam, Christine (ed.), *Dental Practice in Europe at the End of the Eighteenth Century* (Amsterdam: Rodopi, 2003), 171–310.

Harvey, Karen, "The Majesty of the Masculine Form': Multiplicity and Male Bodies in Eighteenth-Century Erotica' in Hitchcock, Tim and Cohen, Michèle (eds), *English Masculinities, 1660–1800* (London: Longman, 1999), 193–213.

Lane, Joan, 'The Diaries and Correspondence of Patients in Eighteenth Century England' in Porter, Roy (ed.), *Patients and Practitioners: Lay perceptions of medicine in pre-industrial society* (Cambridge: CUP, 1985), 205–48.

Lawrence, Ghislaine, 'The Ambiguous Artifact: Surgical Instruments and the Surgical Past' in Lawrence, Christopher (ed.), *Medical theory, surgical practice: studies in the history of surgery* (London; New York: Routledge, 1992), 295–314.

Lawrence, Susan C., 'Private Enterprise and Public Interests: Medical Education and the Apothecaries' Act, 1780–1825' in French, Roger and Wear, Andrew *British Medicine in an Age of Reform* (London: Routledge, 1991), 45–73.

McKendrick, Neil, 'The Consumer Revolution of Eighteenth-Century England' in McKendrick, Neil, Brewer, John and Plumb, J.H. (eds), *The Birth of a Consumer Society* (Bloomington: Indiana University Press, 1982), 9–33.

McKendrick, Neil 'George Packwood and the Commercialization of Shaving: The Art of Eighteenth-Century Advertising or "the Way to Get Money and Be Happy"' in McKendrick, Neil, Brewer, John and Plumb, J.H. (eds), *The Birth of a Consumer Society* (London: Europa Publications, 1982), 146–96.

Payne, Lynda, 'Drain, Blister, Bleed: Surgeons Open the Skin in Georgian England' in Reinarz, Jonathan and Siena, Kevin (eds), *A Medical History of Skin* (London: Pickering and Chatto, 2013).

Pelling. Margaret, "Medical Practice in Early Modern England: Trade or Profession?" in Prest, Wilfrid Robertson (ed.), *The Professions in Early Modern England* (London: Routledge, 1987), 90–128.

Perez, Liliane Hilaire, 'Technology Curiosity and Utility in France and England in the Eighteenth Century' in Bensaude-Vincent,

DOI: 10.1057/9781137467485.0012

Bernadette and Blondel, Christine (eds), *Science and Spectacle in the European Enlightenment* (Farnham: Ashgate, 2008), 25–42.

Pointon, Marcia, 'Jewellery in Eighteenth-Century England' in Berg, Maxine and Clifford, Helen (eds), *Consumers and Luxury: Consumer Culture in Europe 1650–1850* (Manchester: Manchester University Press, 1999), 120–46.

Schaffer, Simon, 'Enlightened Automata' in Clark, William, Golinski, Jan and Schaffer, Simon (eds), *The Sciences in Enlightened Europe* (Chicago: University of Chicago Press, 1999).

Turner, David M., 'The Body Beautiful' in Reeves, Carole (ed.), *A Cultural History of the Human Body in the Enlightenment* (London: Bloomsbury, 2010), 113–32.

Vila, Anne, 'Medicine and the Body in the French Enlightenment' in Brewer, Daniel (ed.), *The Cambridge Companion to the French Enlightenment* (Cambridge: CUP, 2014), 199–213

Journals

Berg, Maxine and Clifford, Helen, 'Selling Consumption in the Eighteenth Century: Advertising and the Trade Card in Britain and France', *Cultural and Social History*, 4:2 (2007), 145–70.

Berry, Helen, 'Polite Consumption: Shopping in Eighteenth-Century England', *Transactions of the Royal Historical Society*, 12 (2002), 375–94.

Bloomfield, Anne and Watts, Ruth, 'Pedagogue of the Dance: the Dancing Master as Educator in the Long Eighteenth Century', *History of Education*, 37:4 (2008), 605–18.

Booth, Christopher, 'The Rod of Aesculapios: John Haygarth (1740–1827) and Perkins' Metallic Tractors', *Journal of Medical Biography*, 13:3 (2005), 155–61.

Cohen, Michèle, '"Manners" Make the Man: Politeness, Chivalry and the Construction of Masculinity, 1750–1830', *Journal of British Studies*, 44:2 (April 2005), 312–329.

Coppola, Al, '"Without the Help of Glasses": The Anthropocentric Spectacle of Nehemiah Grew's Botany', *The Eighteenth Century*, 54:2 (2013), 263–77.

Corfield, Penelope, 'Dress for Deference and Dissent: Hats and the Decline of Hat Honour', *Costume*, 23 (1989), 64–79.

Craddock, Paul., and Lang, Janet, 'Crucible Steel – Bright Steel', *Historical Metallurgy*, 38:1 (2004), 35–46.

DOI: 10.1057/9781137467485.0012

Deutsch, Phyllis, 'Moral Trespass in Georgian London: Gaming, Gender and Electoral Politics in the Age of George II', *Historical Journal,* 39:3 (1996), 637–56.

Ellis, Markman, 'Coffee-House Libraries in Mid-Eighteenth-Century London', *The Library,* 10:1 (2009), 3–40.

Evans, Chris, 'Crucible Steel as an Enlightened Material', *Technology and Culture,* 42:2 (2008), 79–88.

Evans, Chris and Withey, Alun, 'An Enlightenment in Steel?: Innovation in the Steel Trades of Eighteenth-Century Britain', *Technology and Culture,* 53:3 (2012), 533–60.

Festa, Lynn, 'Cosmetic Differences: The Changing Faces of England and France' *Studies in Eighteenth-Century Culture,* 34 (2005), 25–54.

Finn, Margot, 'Men's Things: Masculine Possession in the Consumer Revolution', *Social History,* 25:2 (2000), 133–155.

Fisher, Will, 'The Renaissance Beard: Masculinity in Early Modern England', *Renaissance Quarterly,* 54:2 (2001), 155–87.

Flick, C.S., 'Spectacles as Badges of Rank and Learning', *Vision,* 3:1 (1949), 36.

Gillespie, Neal C., 'Natural History, Natural Theology and Social Order: John Ray and the "Newtonian Ideology"', *Journal of the History of Biology,* 20:1 (1987), 1–49.

Greenblatt, Stephen, 'Filthy Rites', *Daedalus* 111:3 (1982), 1–16.

Jones, Colin, 'The Great Chain of Buying: Medical Advertisement, the Bourgeois Public Sphere, and the Origins of the French Revolution', *The American Historical Journal,* 101:1 (1996), 13–40.

Jones, M., 'Industrial Enlightenment in Practice: Visitors to the Soho Manufactory, 1765–1820', *Midland History,* 33:1 (2008), 68–96.

Kirkup, John, 'From Flint to Stainless Steel: Observations on Surgical Instrument Composition', *Annals of the Royal College of Surgeons of England,* 75 (1993), pp. 365–374.

Klein, Lawrence, 'Politeness and the Interpretation of the British Eighteenth Century', *Historical Journal,* 45:4 (2002), 869–98.

Kook, N.J., 'Metal Frame Materials', *MOI Symposium Paper,* March 1978, 17–18.

Kwass, Michael, 'Big Hair: A Wig History of Consumption in Eighteenth-Century France', *American Historical Review,* 111:3 (2006), 631–59.

Martin, Morag, 'Doctoring Beauty: The Medical Control of Women's *Toilettes* in France, 1750–1820, *Medical History,* 49:3 (2005), 351–68.

DOI: 10.1057/9781137467485.0012

Oldstone-Moore, Christopher, 'The Beard Movement in Victorian Britain', *Victorian Studies* 48:1 (2005), 7–34.

Perez, Liliane Hilaire and Rabier, Christelle, 'Self Machinery? Steel Trusses and the Management of Ruptures in Eighteenth-Century Europe', *Technology and Culture*, 54 (2013), 460–502.

Phibbs, John, 'The View-Point', *Garden History*, 36:2 (2008), 215–17.

Pincus, Steven, '"Coffee Politicians Does Create": Coffee-Houses and Restoration Political Culture', *Journal of Modern History*, 67 (1995), 807–34.

Picciotto, 'Optical Instruments and the Eighteenth-Century Observer', *Studies in Eighteenth-Century Culture*, 29 (2000), 123–53.

Powell , Margaret K. and Roach, Joseph R. 'Big Hair', *Eighteenth-Century Studies*, 38:1 (2004), 79–99.

Riskin, Jessica, 'The Defecating Duck, or, the Ambiguous Origins of Artificial Life', *Critical Inquiry*, 29:4 (2003), 599–633.

Rosenthal, Angela, 'Raising Hair', *Eighteenth-Century Studies*, 38:1 (2004), 1–16.

Stelmackowich, Cyndy, 'The Instructive Corpse: Dissection, Anatomical Apecimens and Illustration in Nineteenth-Century Medical Education', *Spontaneous Generations: A Journal for the History and Philosophy of Science*, 6:1 (2012), 50–64.

Stobart, Jon, 'Selling (Through) Politeness: Advertising Provincial Shops in Eighteenth-Century England', *Cultural and Social History*, 5:2 (2008), 309–28.

Styles, John, 'Product Innovation in Early Modern London', *Past and Present*, 168:1 (2000), 124–169.

Talbot, Stuart, 'Shagreen and Fish Skin: On Scientific Instruments, Their Cases and Etui, circa 1700–1800', *Bulletin of the Scientific Instrument Society*, 123 (2014), 10–27.

Tosh, John, 'What Should Historians Do With Masculinity?: Reflections on Nineteenth-Century Britain', *History Workshop*, 38 (1994), 179–202.

Turner, David M. and Withey, Alun, 'Technologies of the Body: Polite Consumption and the Correction of Deformity in Eighteenth-Century England', *History*, 99:338 (2014), 775–96.

Von Rohr, Moritz and Court, Thomas H., 'Contributions to the History of the Spectacle Trade from the Earliest Times to Thomas Young's Appearance', *Transactions of the Optical Society*, 25:2 (1923–4), 41–72.

Vickery, Amanda, 'His and Hers: Gender, Consumption and Household Accounting in Eighteenth-Century England', *Past and Present*, Supplement 1 (2006), 12–38.

DOI: 10.1057/9781137467485.0012

Wahrman, Dror, 'Change and the Corporeal in Seventeenth and Eighteenth-Century Gender History: Or, Can Cultural History be Vigorous?', *Gender and History,* 20:3 (2008), 584–602.

Williams, Carolyn D., '"Half a Charge and No Wadding": Women and Guns in the Eighteenth Century', *Journal for Eighteenth-Century Studies,* 25:2 (September 2002), 247–65.

Wrigglesworth, Jeffrey R., 'Bipartisan Politics and Practical Knowledge: Advertising of Public Science in Two London Newspapers, 1695–1720', *British Journal for the History of Science,* 41:4 (2008), 517–540.

Theses

Smith, Virginia, 'Cleanliness: Ideas and Practice in Britain, c. 1770–1850', Unpublished PhD diss., London School of Economics., 1985.

DOI: 10.1057/9781137467485.0012

Index

DOI: 10.1057/9781137467485.0013

DOI: 10.1057/9781137467485.0013

DOI: 10.1057/9781137467485.0013

DOI: 10.1057/9781137467485.0013

DOI: 10.1057/9781137467485.0013

Lightning Source UK Ltd.
Milton Keynes UK
UKOW04n2033250216

269102UK00002B/6/P